SMOKIN'
JOE

SMOKIN' JOE

THE LIFE OF JOE FRAZIER

MARK KRAM, JR.

ecco
An Imprint of HarperCollins*Publishers*

SMOKIN' JOE. Copyright © 2019 by Mark Kram, Jr. All rights reserved. Printed in the United States of America. No part of this book may be used or reproduced in any manner whatsoever without written permission except in the case of brief quotations embodied in critical articles and reviews. For information address HarperCollins Publishers, 195 Broadway, New York, NY 10007.

HarperCollins books may be purchased for educational, business, or sales promotional use. For information please email the Special Markets Department at SPsales@harpercollins.com.

FIRST EDITION

Designed by Suet Chong

Library of Congress Cataloging-in-Publication Data has been applied for.

ISBN 978-0-06-265446-5

19 20 21 22 23 LSC 10 9 8 7 6 5 4 3 2 1

FOR CORY AND OLIVIA

A CHAMPION IS SOMEONE WHO GETS UP WHEN HE CAN'T.
—JACK DEMPSEY

CONTENTS

THE LOVE

Joe Frazier, 1968.
AP Images

H is straw fedora tipped at a jaunty angle, Joe Frazier shuffled through the door of his upscale apartment with a pocketful of lottery tickets and something for Denise, his loyal companion of forty-one sporadically stormy years. Whenever he would go out to play a number at a convenience store—typically ten dollars a throw, now and then more if he had a hunch the stars were favorably aligned—he would spot an item as he strolled through the aisles, impulsively buy it in bulk, and bring it back for her as a small acknowledgment, the way an appreciative house cat drops an expired mouse at the feet of its owner. A week before, he had exited the elevator behind a handcart bearing eight cases of soda pop. Today, he showed up with enough paper towels under his arm to sop up an oil spill. "Look what I got for you!" he said, his weathered face unfolding into a wide smile. Forever amused by these impromptu deliveries, Denise Menz would think of how she and Joe always had paper towels with them whenever they crisscrossed America by car. Afraid of planes but not of weaving in and out of tractor trailers on the highway at upwards of 110 mph, he would use them to swab himself down with rubbing alcohol instead of stopping at a hotel to take a shower. Denise would later say, "Funny what you would remember."

The apartment was on the twentieth floor of a building that overlooked the Ben Franklin Parkway, at the far end of which the Rocky statue stood sentry at the Philadelphia Museum of Art. With her background as an interior designer, Denise had given the space an eclectic look that embraced a masculine color scheme. While Joe found it amenable enough, he yearned for the five-thousand-square-foot loft above his North Philadelphia gym, which Denise had decorated in the style of a 1970s bachelor pad and had outfitted with accents that included a leopard-spotted chaise lounge in the bathroom. For the better part of three decades, he had holed up amid the consoling shadows of 2917 North Broad Street, the very place where he had whipped his body into shape for his epic showdowns with Muhammad Ali and where years later he would still don an old robe and pad downstairs to paw at the heavy bag. But Joe was in uncertain health, the abode was cold and damp, and Denise and others were concerned that he could no longer climb the steep stairs to his quarters without falling. Plus, there were unending tax hassles. So Denise had procured the spot in Center City where the two now lived.

By virtue of a reportorial acquaintance with Joe that dated back to my father, Mark Kram, who covered him for *Sports Illustrated* from the early days of his career, I stopped in to see him one day in June 2009 for a piece I was thinking of doing for an overseas magazine. At sixty-five, his handshake was still firm, scarcely the grip of a man rumored to be in declining health. Word was that he was battling diabetes and high blood pressure, and that he still had not recovered physically from a car crash in 2002. Surgeries had followed on his back and neck, yet he remained in some degree of pain, which his eyes betrayed as he lowered himself to sit. Even so, he appeared full of cheerful contentment, far removed from the enduring portrayal of him as an angry and unforgiving man so incapable of letting go of the hatred he harbored for Ali. That Joe was elsewhere on this spring day.

Joe relaxed at his dining room table and picked at a bowl of cherries. As Denise checked on the ribs she had in the oven, Joe and I found ourselves on the subject of his old R&B group. With Joe as the lead vocalist and principle financier, Smokin' Joe and the Knockouts played the club circuit back in the 1970s. Overhearing this turn in the conversation, Denise chimed in: "Joe had a deal with Capitol. He cut some forty-fives. We have a stack of them around here somewhere." The act never climbed to the heights Joe had hoped, yet the songbird in him was still apt to soar with unbidden spontaneity, even given the presence of an audience of just himself. With a raspy voice, he launched into a rendition of "My Way," the lyrics of which had been personalized for him.

I've come a long, long ways
And like they say
It took some doin' . . .
I fought them fair
I fought them square
I fought them my way. . . .

"Remember that one—'My Way'?" Joe asked, his eyes crinkled with merriment. "Paul Anka rewrote some of the words to it just for me."

His way was the hard way. In the ring, he lived and died by the simple yet daring principle of engagement that in order to deliver one bone-crunching blow, it was frequently necessary to absorb three in exchange. With a left hook that was by acclamation an instrument of doom, he would leave behind a crimson trail of swollen eyes and broken jaws in his quest for his place in history. One opponent would say that the volume of punches Frazier had battered him with was so unrelenting that it was "like getting hit by four hands." When another was revived back in his dressing room, he began tying his shoes on again until one of his handlers had to break it to him that he had already been knocked flat. Even Ali would say of his war with Frazier in Manila: "It was like death. Closest thing to dying I know of." For his legions of fans, he possessed the raw power of Rocky Marciano—the Rock—who had set the standard for perfection when he retired with a 49-0 record in 1956; others claimed he swarmed his opponents with the intensity of Henry Armstrong, who fought 181 bouts in the 1930s and '40s and won championships as a featherweight, lightweight, and welterweight. Whoever one supposed his antecedents were, Joe exuded an atavistic joy inside the velvet ropes that was his and his alone. Only one outcome seemed certain whenever Joe charged into action: someone would end up in the emergency room.

On his excursions across Philadelphia in later years, he had a prepared answer for anyone who asked how he was doing. With an air of playful exasperation, he would shake his head sadly and reply with a shrug, "They're tryin' to get me!" No one ever knew to whom he was referring, but it could have been anyone—the taxman, a woman he had been with the evening before, or perhaps some phantom of a bigoted world that was never far away. He had grown up in Jim Crow South Carolina, where African Americans remained entrapped in a plantation culture that subjected them to indignities not so far removed from the experience of their enslaved ancestors. When he came to the North as a young teenager, he endured degradations no less oppressive than the ones he had encountered back home in Beaufort. Even as he came into his own as a professional athlete in the 1960s and '70s, which afforded him the proximity to white money and the sanctuary it provided, he came face-to-face with "black-on-black" hate language in his exchanges with Ali. From cradle to grave, it would be a journey galvanized by conflict.

Unimposing for a heavyweight, at just under six feet tall, he was a beloved overachiever who once said of himself, "I'm a small piece of leather, but I'm well put together." In an era that would come to be looked upon as the golden age of the heavyweights, Joe did not back up an inch as he battled his way toward a place at the top. Even as his opponents pummeled his head again and again, he advanced upon them with a gallantry that thrilled crowds. Fans could see themselves in his clock-in-early, leave-late work ethic. While he did not possess the height or personality of some of his peers, no one would ever have cause to question the size of his heart or his courage under fire. Few men in the annals of the ring produced moments as indelible.

Far more than just an appealing athletic proposition, the rivalry that commenced between Frazier and Ali on March 8, 1971, was a cultural happening that exposed the deep fissures in American society. By an accident of circumstances, they ended up in the crosshairs of an argument far larger than themselves. Ali: exiled from the ring for his evasion of the draft during the Vietnam War and scorned for his conversion to the Nation of Islam. Frazier: up from the abject poverty of the South Carolina Lowcountry and later a dead-end job in a Philadelphia slaughterhouse, the involuntary hero to that sector of society in lockstep with the "white chauvinist" ethic of TV character Archie Bunker. By the end of their trilogy four years later, we would learn as much about ourselves as we would about either of them.

For forty-one rounds of concentrated, pedal-to-the-floor action, they pushed each other to the very edge of human endurance. Frazier won their initial encounter at Madison Square Garden, the celebrity extravaganza that was billed as "The Fight of the Century." Ali claimed their January 1974 rematch at the Garden in a somewhat less compelling tactical effort. Then there was the final act: Manila. On a scorching morning in October 1975, Ali prevailed in their rubber match when Frazier, battered and unable to see, was stopped by his cornerman from answering the bell for the fifteenth and final round. For his strivings to end in capitulation was no more agreeable to Frazier than asking him to swallow a cup of lye, particularly in light of the rumors that would persist for close to forty years that Ali himself had been on the verge of quitting.

Whatever heights of athletic achievement they drove each other to inside the ring, they dragged each other down in a running feud outside of it. For Ali, it began as a showy ploy to draw attention to himself and pump the gate. Even as he poked at his opponents in disparaging ways and could be boorish in the eyes of his decriers, his act was looked upon by his devotees as the greatest show on earth that did not employ a trapeze artist. Loud, funny, and irrepressibly original, Ali peppered his oratory with amusing doggerel that heightened the fervor of his fans, if not his standing in literary circles. But he crossed a line when it came to Frazier, who grew irritable as Ali turned from playful to hostile. With the cameras rolling, Ali called him "ugly," a "gorilla," and an "Uncle Tom." Frazier shot back and called Ali by his "slave name"—Cassius Clay. Increasingly furious as Ali branded him a pawn of white establishment figures such as Philadelphia mayor Frank Rizzo, Frazier brooded that Ali had turned some portion of the black community against him. Privately, Ali told him that he was only whipping up fan interest that would reap a big payoff for both of them. But it annoyed Joe when Ali settled down in Philadelphia during his exile and began showing him up in public. When Ali heckled him on the street one day in

Center City in 1970, Frazier became so incensed, two of his associates told me that he threatened to take a tire iron to him and finish him off before they ever stepped into the ring.

The hard feelings Ali had stirred in Joe waxed and waned in the years that followed. On the day I was with him at his apartment, he had slipped behind the public face that those closest to him had encouraged him to wear and laughed off "the Butterfly" with an air of letting bygones be bygones. On other days, I was with him when his annoyance spiked like a fever, his voice full of anger and hurt that some say he carried to his grave. As the years unfolded and Ali grew infirm with Parkinson's disease, as his speech became slurred and his hands increasingly quivered, Frazier appeared to take cruel pleasure in the adversity that had befallen "Clay." "Look at him, and now look at me," he told me and others. "Who do you think came out the winner?" Somehow, he had convinced himself that his signature was embossed on the physical wreck Ali had become. Even as his friends reminded him that Ali was a sick man and implored him to back off, Frazier could not help himself from battering his erstwhile rival with verbal haymakers. For those who had followed the dips and curves in the acrimony between the two, Joe seemed like a Japanese soldier on some island in the Pacific who had not yet heard that World War II had ended. Incisively, author David Halberstam observed in an essay, "Technically the loser of two of the three fights, [Frazier] seems not to understand that they ennobled him as much as they did Ali, [and] that the only way we know of Ali's greatness is because of Frazier's equivalent greatness." Whatever truth there is in that, there was no equivalency between them when it came to the accumulation of what Joe called "the love."

"The love" was money. In terms of sheer earning power, Joe was no more in the same league with Ali than he was with Otis Redding as a singer. But Joe did well for himself during and in the aftermath of his career, far better than he could ever have dreamed of when he was a boy helping his father cook moonshine in the woods. By 1970, even before he and Ali each earned a colossal $2.5 million for the Fight of the Century, he had become ensconced in the comforts of the upper middle class. He and his wife, Florence, had five children, all of whom attended good schools; they lived in a $125,000 house on two and a half acres in a fashionable Philadelphia suburb; he had four hundred thousand dollars in cash in the bank and a variety of investments; and he had indulged himself with the acquisition of assorted automobiles, including a gold Cadillac with not one but two telephones; a Corvette Stingray; and a 1934 Chevrolet, with which he was forever tinkering. To the everlasting horror of the consortium of Philadelphia businessmen who backed him, he also owned a Harley-Davidson Electra Glide, which he tricked up and boasted that he "got up to 110 mph" on a straightaway. Forty years later, he was still bringing in a handsome living from appearances and such, yet there would never be enough "love" on hand to offset his swelling obligations to the Internal Revenue Service and others. With the exception of assorted pieces of memorabilia, he died in 2011 with no assets to speak of and only a few dollars in the bank.

Given the penury from which he sprang and the disadvantage of just a ninth-grade education, nothing had prepared Frazier for the vast sums of money that would come into his life or the international fame he would achieve. Early on, his day-to-day affairs were overseen by his investment group, Cloverlay, which saw to it that his taxes were paid. But once he parted ways with them in 1974 at the end of an eight-year run, it was not long before his finances fell into disarray. And as his marriage to Florence unwound over a period of years, he answered the siren call of the opposite sex with alacrity yet endeavored to be a vigilant provider. By Florence and five other women, he had a total of eleven offspring—six daughters and five sons. Whatever ephemeral comfort Joe found in the arms of others, however it eased the pain that had settled into his bones from years of being hit, it came at a price that converged with his yen for gambling and certain investment reversals to leave him in the red. At the suggestion of a confidant that he declare bankruptcy, Joe said flatly, "No, I'll pay back every penny I owe."

Even though he would fight two more times after Manila, clubbed into submission by George Foreman in their 1976 rematch and dragged to a draw five years later by the obese ex-jailbird Floyd "Jumbo" Cummings, he turned with renewed gusto to singing and began managing young boxers at the gym. Chief among them was his son Marvis, who had shown promise as an amateur heavyweight and who had escaped death when he sat out a scheduled trip to Poland in which fourteen American boxers perished in a plane crash. Joe had a premonition and commanded Marvis not to go. But Joe could not save his oldest son from swift annihilation in the ring, which came first at the hands of reigning champion Larry Holmes and later by the buzz saw that was rising star Mike Tyson. By disposition a sweet man who possessed none of the raw killer instinct that Joe had, Marvis became an ordained minister and remained a ubiquitous presence at his father's side. He accompanied him to his appointments and acted as a voice of caution in his ear. When an occasional well-intentioned friend would buttonhole Joe and talk up the possibility of a comeback—which would leave his father with a gleam of temptation in his eye—Marvis would take the friend aside and gently say, "Please, leave Pop be." Joe told me more than once, "Every father should have a son like Marvis."

Big-hearted in ways that the crowds would seldom see, Joe looked upon the work he did with the boys and girls at the gym as more than just boxing instruction. Only a few would have any chance in the sport in which he had distinguished himself, yet he understood that his calling was far larger. For the ones with nowhere to go but the streets when school let out, he provided a place to work off the simmering rage that poverty breeds. For those who came to him with some ability and the same dream he once had, he provided a bed to sleep in at the gym and a couple hundred dollars each week to get by on, the way Cloverlay had given him a leg-up years before. Along with Marvis, he worked with his other sons, daughters, and assorted nephews and nieces, who would come to appreciate and indeed love him not just for his hallowed accomplishments in the ring but also for how he gave of himself to them in large and small ways. As he sat with me that day in his apartment, it did not escape him that the way forward for him had been paved not just by his own hard work but by the help of others.

"I was born into animosity, bigotry, hatred, and white water/colored water," he told me. "I look back on those days and I think: Well, you are a better man because of them. The world has changed to the point where we now have a black president. But the youngsters still need to be shown that someone cares."

Big names would come and go at the gym for years—fellow champions, actors, politicians, and journalists by the score. Whenever Joe would drop by, parking his Cadillac up on the sidewalk by the gym entrance, someone would spot him and shout, "Hey, Champ!" Grinning, Joe would wave and reply, "Yo, man!" In an area of boarded-up houses, he would sit with the firemen at the station around the corner and occasionally play half-ball with them on summer evenings. When he was not working in the gym, chances were that he would be outside under the hood of his car fixing something. He loved tinkering with cars. Whenever he was driving someplace and spotted one with a flat, he pulled over, introduced himself to the distressed driver, and changed the tire. But it was an episode that occurred on a December day in 1986 that revealed to former cruiserweight Kevin Dublin the type of man Joe Frazier was.

Holiday lights blinked in the windows along North Broad Street on that cold, cold day. Joe was on his way to Atlantic City, where one of his fighters—heavyweight Bert Cooper—had a bout that evening at Resorts International. On his way out of the gym door, Joe ordered his son Hector and Dublin to come along, if just to keep an eye on them; both were just starting out in careers of their own. With Joe behind the wheel and Hector and Dublin in the back, the limousine glided down Broad toward the Ben Franklin Bridge, only to draw to a sharp stop as Joe came upon a man with no legs crossing the street in a wheelchair with a can of kerosene in his lap. Joe parked the car at an angle on Broad and hopped out, dressed in a long fur coat and cowboy hat, as passersby stopped on the sidewalk and looked on in curiosity.

"Come on," Joe said to the man in the wheelchair. "Looks like you need some help."

Joe picked up the man and placed him in the passenger seat, as Hector and Dublin stowed the wheelchair and the can of kerosene in the trunk and snapped it shut. Joe asked the man, "Where you headed?" When the man gave him a nearby address, Joe steered the limo up and down some side streets and found it. Leading up to the door was a ramp constructed of haphazard pieces of splintered wood. Joe placed him in the chair that Hector set up and pushed him to the door. Dublin followed behind with the can.

Covering the windows of the tiny, narrow house were quilted blankets to keep the cold out and the heat in. Dublin remembered thinking once he stepped inside that the occupants were squatters, yet he could not be sure. Their few possessions were scattered in the living room and dining room, including a table and chairs, a TV, and two kerosene heaters. In the shadows were three children. A woman came out of the adjoining room and stopped short when she recognized Joe. The woman squealed, "Lord, look who it is!"

Joe looked down at the man and said, "You look like you could use some love."

The man replied, "No, man, you already showed me love by picking me up and bringing me home."

"Nah," said Joe. He stooped over and pulled a roll of hundred-dollar bills out of his sock. He peeled one off, then another. He handed the man the money.

The man looked up at him at with bewilderment in his eyes and asked, "Why you do this?"

Joe replied, "You need some help."

Joe signed some photos that Hector retrieved from the limo and they were on their way to Atlantic City. Ordinarily, whenever Joe was driving anywhere, there would be Bobby Womack or some other soul singer blaring from the speakers. But not today. It was quiet, as Dublin remembered, "Kind of weird." For close to an hour, Joe barreled down the Atlantic City Expressway and said nothing, the trees alongside the highway spinning by. Then, unprompted by any question or comment from the young men in the back, he addressed Hector and Dublin.

"See, that was a man. Going out in this cold weather to get heat for his family."

There was another long silence. And then Joe said, as if to himself—a tear in the corner of his eye—"You never know, man."

CHAPTER ONE

BILLY BOY

School Days 1955-57

Joe at age twelve.
Philadelphia Inquirer

F
ar and wide across Beaufort County, the boy who would one day become heavyweight champion was celebrated for his uncommon strength. Groups of teenagers would stop by the commercial farm where he worked just to watch him load crates of vegetables into the rear of trucks. Chubby yet endowed with broad shoulders and iron biceps, he prided himself on his physical prowess, which he used to protect less able boys from school bullies. At a Frazier summer reunion years later, his aging cousins remembered

how he would show off by scooping up two of them under each arm and three more on his back. As they laughed and laughed, he would set his jaw and wobble this way and that for ten yards or so before setting them down in a squirming three-hundred-pound pile. Whether this particular piece of lore has been colored by exaggeration is unclear, yet there can be no arguing that by adolescence the seed of who Joe Frazier would become had been planted.

Joe Frazier answered to the name "Billy Boy" in those days. It was inadvertently given him by his doting father, Rubin Frazier, a one-armed handyman-cum-bootlegger whose speech impediment prevented him from uttering the words "baby boy." In the years that would follow, he would be identified by a variety of other appellations—the Slaughterhouse Kid; Billy Joe; later, more famously, Smokin' Joe; or—if he was in the company of friends—Smoke. But it was always "Billy" that would anchor him to the hypnotic South Carolina Lowcountry of his youth. Even at the height of his boxing career, whenever a voice would call out "Hey, Billy!" from the rear of the crowd, he'd immediately send someone in search of the voice. "That's somebody from back home," he'd say. "Go get 'em." No one who knew him from the Laurel Bay section of Burton, South Carolina, ever called him Joe, and surely no one who represented themselves as kin and actually were. They called him Billy Boy or Billy.

Laurel Bay sat in the poorest county in the poorest state in America when Joseph Frazier was born, on January 12, 1944. In a clearing cut from ten acres of woodland dense with big pines and gnarled oaks dripping with Spanish moss, Rubin Frazier erected a single-story house that had a sporadically leaky roof and no electricity or running water until some years later. Elevated from the ground on oak blocks, thus keeping it dry in the event of occasional flooding from nearby creeks and swamps, the bare structure had a living room, a kitchen, and two bedrooms—one occupied by Rubin and his wife, Dolly; Billy; and Billy's sisters, Martha (Mazie), Julia

(Flossie), and Rebecca (Bec); and the other variously by his brothers Marion (Bubba), Eugene (Skeet), Andrew (Bozo), Rubin Jr. (Jake), John (Big Boy), and Thomas (Tommy). David, a seventh brother, would die in infancy of diphtheria. With the exception of an old car that Rubin always seemed to be under the hood of, the place appeared to have been preserved in amber from the bygone days of Reconstruction. Rubin buried what small sum of cash and few valuables he had in a coffee can under the pigpen out back. There, he also stashed his reserves of corn liquor, guarded by a hog that only Rubin seemed to have the skills to pacify. "That animal would take you out," Joe's niece Lisa Coakley told me. "No one would ever get near it but Uncle Billy. It stunk back there."

Generations of Fraziers had labored in the blazing heat on this very same land, first as slaves on the antebellum plantations strewn across the Sea Islands and later as field workers for truck farms that became the hub of the local economy. Slavery had been established at the outset of the founding of South Carolina by prosperous white planters from the West Indian colony of Barbados. The Barbadians had amassed their wealth through the harvesting and processing of sugarcane, which they sold at a premium as molasses, refined white sugar, and rum. Liberal with the whip and other forms of torture, the Barbadians were particularly cruel to their slaves. Because Barbados was relatively small and lacking in economic potential, the planters set sail for the virgin coast of South Carolina, where the chief cash crops were rice, indigo, and, later, cotton. Accordingly, they enslaved specific West African nationalities who were able to endure the extreme climate of the Deep South. Although the identity of their African forebears is uncertain due to the incineration of vital records during the Civil War, DNA analysis indicates that the Fraziers originally came from what are now Cameroon/Congo, Benin/Togo, Mali, Nigeria, and Senegal.

Sea Island slaves were freed early in the Civil War. Strategically advantageous because it had the deepest harbor south of the

Chesapeake Bay, Beaufort was captured from the Confederacy in the Battle of Port Royal, in November 1861, and became a beachhead for Union naval operations. Thus, the ten thousand enslaved Africans who had been in captivity in the area became the initial beneficiaries of Reconstruction. Chief among the progressive initiatives ordered from Washington was the Port Royal Experiment, which established schools for former slaves and enabled them to purchase parcels of the land that had been abandoned by fleeing Confederates. As Reconstruction unfolded in the years that followed, Beaufort enjoyed a flourishing phosphate industry and an international demand for Sea Island cotton. So-called one-mule farmers could earn three hundred dollars a year selling cotton.

Prosperity ground to a halt in the 1890s. By happenstance, the phosphate plants were wiped out during the savage hurricane of 1893 just as prices on the cotton exchange plunged. Amid the wreckage of the hurricane, Beaufort County was deprived of aid from the state legislature, which dismissed it with the same contempt that Governor and, later, Senator Benjamin "Pitchfork Ben" Tillman exhibited when he called it "the niggerdom of Beaufort." By 1895, the state legislature had done away with the Reconstruction constitution and replaced it with a Jim Crow constitution, which placed constraints on the ability of the "colored" population to vote and institutionalized separate entrances, waiting rooms, and drinking fountains in public places. But Beaufort County remained tethered to the Republican Party—the Party of Lincoln—in large part due to the presence of Robert Smalls, an African American who commandeered a Confederate transport ship in 1862 and later became a Republican icon as a U.S. congressman. Unseated from his position in 1886, the so-called King of Beaufort County became port collector and represented the lingering vestiges of the Republican voting bloc until his passing in 1915. Local historian Lawrence Rowland told me: "You could say Reconstruction of the South began in Beaufort County and ended there."

Over the fifty-year period from 1890 to 1940, 52 percent of the African Americans living in Beaufort County joined the Great Migration to the urban Northeast, Midwest, and West. Opportunity presented itself in those places, even if the level of bigotry was no less corrosive than it was in the South. At the height of the Great Depression, in the 1930s, the average annual salary in Beaufort County was ninety dollars—$1.73 a week. On top of that, there were a plethora of diseases. Outhouses became a haven for the parasites that caused hookworm. Malaria was so commonplace that Rowland said "nearly everyone had a touch of it." Vitamin deficiencies led to recurring outbreaks of pellagra and scurvy. Fifty percent of African American males suffered from syphilis, which rendered them ineligible for service in World War II. In preparation for the three-volume history he coauthored on Beaufort County, Rowland found two cases of people who succumbed to starvation. One occurred very close to where the Fraziers lived.

But character, ingenuity, and hard work always provided a way to get by in the Sea Islands. Rubin and Dolly Frazier possessed each in abundance and encouraged these principles in their offspring, who were as hard-nosed outside the ring as Billy would one day become inside it. To feed themselves, the Fraziers hunted and fished, and they sold whatever they could for extra money, be it firewood, manure, or corn liquor. To heal themselves, they combed the tangled brush for "root," plant life that could be distilled into home remedies to be applied topically or swallowed by the spoonful. And if the privation visited upon them became too unbearable, they offered it up to the Lord in prayer, certain that there would be better days ahead. Even if they did not live to see them, it was their abiding belief that their children would, if not here on earth, then at the throne of God.

No one would be able to say years later exactly what precipitated it. The version that Joe told in his autobiography was that it began as an argument between rivals over a woman at a house party. But the account passed down to Lisa Coakley by her grandmother Dolly was that the quarrel started not over a contested paramour but as a turf war between Rubin and Arthur Smith in the illegal commerce of corn liquor. By wide agreement, Rubin had the preferred stock in the county. Unlike the cloudy swill that some bootleggers circulated through an old car radiator during the distilling process, which introduced toxic traces of antifreeze, the goods Rubin sold were always clear in appearance and posed no threat of poisoning. According to Lisa, Smith had a few drinks too many and grew irritated by something the equally inebriated Rubin said or did. Words were exchanged.

"Rubin, I think we need to leave," said Dolly, the two now sitting in his pickup—Rubin with his right hand on top of the steering wheel, Dolly in the passenger seat, their daughter Rebecca cradled in her left arm as she nursed her. Dolly had her right foot tucked up under her.

"Ah, Dee, don't worry about a thing. I'm all right.'"

"No, I'm just getting this funny feeling we need to leave."

That very instant buckshot began to fly. Wielding a sawed-off shotgun with wild-eyed fury, Smith had fired through the passenger side and windshield. The fusillade struck Rubin in his lower left forearm and shattered the bones in his wrist and hand. Dolly covered up to protect her baby, Rebecca, who would not suffer any wounds. But Dolly was hit by pellets in her left foot. Smith was arrested by police and charged with intent to kill; he later served two years in prison. Rubin was admitted to the hospital, where his forearm was amputated. Doctors also treated Dolly, but fragments of buckshot would remain embedded in her foot. At sudden changes in the weather, she would complain: "Oh, that foot. That bullet is in there moving around."

Dolly was pregnant with Billy when that incident occurred, in June 1943. Only five years old when she began toiling in the fields of white men, she would pass down a fighting spirit to her son that was galvanized by the hard and violent days she had seen. She had been raped as a teenager. But she overcame whatever obstacles she encountered with the help of Scripture and a profound integrity that would not allow her to capitulate to any circumstance that she looked upon as degrading. Given the choice between picking vegetables from sunup to sundown and working as a housekeeper for the wife of a wealthy white man, she could not bring herself to do the latter, even if labor as a field hand was just seasonal. "Very spirited—and very spiritual," neighbor Kenneth Doe told me. "She carried herself in a way that you *had* to respect." Along with her essential generosity of heart, Billy would inherit from Dolly Frazier her work ethic, occasional sharp temper, and general distrust of outsiders. She was not a woman to defy, as her progeny would discover whenever they stepped out of line. In a corner of her spotless house was a switch she had braided from supple branches. She would either use that or send the offending child outside to retrieve a suitable stick—and then let him or her wait . . . and wait . . . and wait until she summoned them for punishment. And when Rubin showed up at the door at the end of one of his weekend sprees? His granddaughter Dannette Frazier told me with a laugh, "Oh, Lord, that woman would kick him into the fireplace."

Dolly and Rubin exchanged marriage vows in 1930. By legitimate occupation, Rubin was a carpenter who did odd jobs for people and later worked as an overseer at the Bellamy Farm. One of thirteen children born to Dennis and Susan Frazier (a midwife who helped deliver Billy), Rubin held a certain stature in the community beyond his aptitude for making corn liquor. "He was what you would call a 'fixer,'" his grandson Rodney Frazier told me. "He helped people in any way he could. And if somebody got in trouble, he would find a way to get them out of town." Even with just one hand, Rubin could still change gears, hammer a nail, and tie his shoes. And he still had a wandering eye for women. In his autobiography, Frazier recounted how Rubin had once told him that his sexual adventures produced a total of twenty-six children, and that the accommodating Dolly accepted them as her own whenever they appeared at the door. Only Rubin himself could say if there was any truth to that pretention of fecundity—or how truly accepting of it his wife was. And yet these very same roving appetites would appear in Billy at a young age, along with the same innate enthusiasm Rubin possessed for lending a helping hand to others.

Far from the pitched battles in Europe and the South Pacific, Laurel Bay was enveloped in a winter chill on the day Billy was born. Of that blessed event, Frazier would say years later, "When I was born, three hundred people gathered 'round the house to see if I'd been born with one arm or not, because my daddy had lost his arm in an accident." With a chuckle, sister Mazie wondered aloud, "Was it three hundred? I don't know. I do know there were a lot of people who wondered if he would have one hand." Recollections of those early days would be preserved by Mazie in a composition book: *"Well, it was January 1944 and I had wanted a baby sister. But the day started, the sun came up and news spread that Dolly Frazier had given birth to a son. . . . When my mother went back to work six months later, my dad became his babysitter and was with him the whole day. Billy Boy would sit in his lap when dad drove the car and pretend to steer it. Dad became his hero. You would have had to have been there to see the love they had for each other. Dad would hand him to me and I became his overseer. I had to feed him and see that he got dressed. He depended on me. I looked out for him. And when he became a young man, he looked out for me."* Years later, Mazie told me, "Wherever I went, he went. Everyone thought he was my child."

A used school bus from G. W. Trask & Sons would take Dolly and the other workers from Laurel Bay and drop them off each morning at the fields they were scheduled to pick. George W. Trask had relocated to Beaufort County from Wilmington, North Carolina, in 1908, followed in 1920 by his son Neil (who later branched off into cattle in Calhoun Falls, South Carolina) and in 1934 by his son John. Shrewdly, John began acquiring land at Depression prices and cultivated key relationships with two top grocery chains, A&P and Safeway. At the height of its operations in the 1940s and '50s, G. W. Trask produced approximately one thousand acres of crops annually, the profits of which were shared with John and his young brother Harold (also known as Beanie). Some years were better than others, depending upon any wild swings in the weather, but it was, by and large, a lucrative undertaking that enabled John to follow his entrepreneurial spirit and buy into a variety of downtown businesses. John Trask also became an influential figure in local politics.

Commercial farms were the chief employers in Beaufort County during these lean years. At G. W. Trask & Sons, field hands numbered anywhere from thirty-five to 150 during any given season. But while work in the fields could be counted on to provide a steady, if small, income, it was a hard existence for the men and women who stooped for long hours over rows of beets, radishes, lettuce, tomatoes, and such. "It was backbreaking labor," John Trask III told me. "You were bending over all day long." Dolly wore a wide-brimmed hat, a scarf, long-sleeved shirts, pants, and work boots. With each bushel she picked, she would be given a "babbitt" valued at twenty-five to fifty cents, depending on the effort involved in picking the specific crop; they were redeemed for cash each Saturday. Annie Green worked with her cousin Dolly in the fields as a young girl and told me, "She had very fast hands." Occasionally, Dolly hired on as a day worker canning tomatoes, which had to be pulled off a conveyor belt and skinned. Her daughter Mazie remembered she would peel and devein shrimp when the trawlers just in from the Atlantic Ocean unloaded their catch at the city dock. The jagged shells would leave her fingers bloody. Dolly once said, "Why, I done ever' little thing for a livin' 'cept kill a man."

Grocery outlays were offset by how far a dollar could go in

those days: a twenty-four-ounce loaf of Marvel Bread cost eleven cents, a pound of cheese thirty-seven cents, a dozen oranges twenty-five cents, and so on. But the Fraziers became skilled at living off the land. On their own ten acres in Laurel Bay, they coaxed peas, cotton, and watermelon out of the sandy soil. Eggs were procured from the chickens that were always underfoot in the yard. At the end of their productive peak, the chickens were butchered, as were the occasional cow and the hogs that were not otherwise engaged in standing sentry over the household loot. Whenever the larder grew bare, the Broad River and its tributaries yielded crabs and oysters and an abundance of fish—whiting, trout, flounder, and spottail bass. "If we had something that someone else needed, we traded it for something we needed," said Mazie. "That was how we got by." Dolly relaxed at the end of the day by lighting up a bowl of tobacco in her corncob pipe.

Forays deep into the woods provided access to a bounty of deer, ducks, turkeys, and small game such as fox, rabbit, raccoon, and squirrel. "The Lord make a way for you there," Dolly would say of the wild, where she used to pick huckleberries to be sold by the quart door-to-door in Laurel Bay and "musk" to be peddled by the pound to "a white man who come around in his truck." Billy would accompany her on these expeditions and help carry out sacks of the latter, yet he would always be spooked by the woods, how dark they would get at night, the animal life that lurked in the jungle of vines, and the noises—the sudden, creepy calls of peril. Instead of venturing outside at night to use the privy, which stood seventy-five yards from the back door, Billy would reach for the "slop jar" that was kept underneath his bed. "When I die," he'd say years later, "don't go buryin' me down in the sticks, ya hear me?" But a cornucopia of "roots" bloomed there that were used as pharmaceuticals to treat fevers (horehound root); skin ailments (cockleburs); coughing (cockroach tree); urination issues (diluted wild grapevine sap); cataracts (persimmon sapwood), and so on. "Cures" were doled out not by medically trained professionals but by "root doctors"—latter-day African "witch doctors"—who in Gullah custom conjured up potions that both addressed physical needs and summoned the assistance of good and evil spirits. They assumed animal names—Dr. Bug, Dr. Buzzard, Dr. Crow, and such—and engaged in an enduring fox-and-hound chase with Ed McTeer, the so-called High Sheriff of the Lowcountry, who himself claimed to be skilled in root yet was obliged by law to haul the root doctors in for practicing medicine without a license.

Visitors would come by with some regularity in search of Stepney Robinson—aka Dr. Buzzard—who was located on St. Helena Island and was renowned for dispensing root in the form of charms intended to be worn, chewed, or buried. "People would come looking for him to get revenge on a rival by casting spells on them," said Mazie. Closer to Laurel Bay was Peter Murray—aka Dr. Bug. Like Dr. Buzzard and the others, he kept a low profile in order to stay a step ahead of the law. An inhabitant of Laurel Bay could earn occasional dollar tips by helping visitors find Dr. Bug, whose career in voodoo came to an end when he began handing out a potion that enabled draftees to evade the service by giving them "hippity-hoppity hearts." When two Gullah draftees died en route to their physicals at Fort Jackson, an investigation by McTeer uncovered that the potion contained corn liquor and a small amount of lead arsenate. "No, that potion ain't pizen," Dr. Bug explained to the court sorrowfully. "I drinks a shot of it myself ever' day." Dr. Bug pleaded guilty to hindering the Selective Service, was freed on five thousand dollars bond, and died soon thereafter.

Given her strong Baptist leanings, Dolly Frazier steered clear of the conjurings of Dr. Bug and his ilk, yet as Joe would remember, she was not beyond scaring you "practically out of your britches" with tales of the dead moaning in their graves. Nor was Dolly enthused by the perils the bootlegging operation invited into her house—certain arrest, possible jail time, the presence of firearms, and the corn liquor itself, which loosened not only tongues but also trousers. But Dolly adopted a pragmatic view that allowed her to accept it: The extra money always helped. Thus, it became a family operation, with Dolly and her daughters pitching in with sales and the sons helping Rubin on the production end. The still was hidden deep in the woods, and they scheduled their work in the early hours of the day, when the landscape would be covered with a ground fog that obscured the smoke curling up from the fire beneath the boiler. To ensure himself a jump on the law, Rubin would get up at 4 A.M., lean over his youngest son as he slept, and whisper in his ear: "Wake up, Billy Boy."

Until he showed up for an appearance at Robert Smalls High School, Joe Louis had been a figure that the black community of Beaufort looked upon only from afar, his storied exploits burnished by his victory over Nazi propaganda tool Max Schmeling in 1938. They'd seen him in grainy newsreels at the downtown Breeze Theater, where blacks had their own entrance and were herded into the balcony, or perhaps in the pages of the *Beaufort Gazette*, which typically confined its coverage of local blacks to a small inside column under the rubric NEWS OF INTEREST TO THE COLORED COMMUNITY. But it was not until the Brown Bomber stepped into the ring in Beaufort to box in an exhibition on March 10, 1950, that he emerged before them in the flesh, his sloping, coffee-colored shoulders draped in a robe. By bus, by car, and by foot, his fans came from across the county and beyond to pay homage to a man who, in the words of novelist Richard Wright, embodied "the concentrated essence of black triumph over white."

Louis had been invited to Robert Smalls by its forward-thinking principal, W. Kent Alston, who in later years would bring in Marian Anderson and Count Basie to perform. Far from the unmatched talent he had once been, Louis had hung up his gloves the year before, only to agree to a comeback bout against champion Ezzard Charles in September 1950 to chisel away at a five-hundred-thousand-dollar debt he owed the Internal Revenue Service. His visit to Beaufort was part of a sixteen-stop exhibition tour that would take him through Georgia, Alabama, Mississippi, South Carolina, and Texas. Given the key to the city in a ceremony presided over by Beaufort Mayor Angus D. Fordham, Louis boxed a scheduled three rounds against an undistinguished local and collected half the $1,290.67 in proceeds. Ever magnanimous, he later said of his opponent John Shaw: "He packs a real punch. I was saved by the bell in that last round."

In the crowd that evening was Billy Frazier. Far too young at six years old to grasp the social significance of the man he had come to see, he was just old enough for the event to leave an impression on him that would take shape in his imagination. "Every Negro boy old enough to walk wanted to be the next Brown Bomber," said Malcolm X, who as a teenager gave boxing a whirl himself. From a knotty oak tree in his backyard, Billy hung a burlap sack that he jammed with rags, sand, corncobs, Spanish moss, and bricks. With his hands wrapped in socks, he would spend an hour each day attacking it with wild blows, hearing in his head the prophecy of his uncle Israel: "That boy's gonna be another Joe Louis!" Dolly's nephew Charles Middleton remembered seeing Billy swat at the bag. "Sometimes it would swing back and knock him in the nose!" he told me. "But he would keep at it." By the early 1950s, when Rubin had arranged to hook up his house to the electrical grid, he purchased a black-and-white television set that acquainted his youngest son with the royalty of boxing in almost nightly telecasts. From Madison Square Garden in New York City, the *Gillette Cavalcade of Sports* aired on Friday evenings and featured showstoppers such as Sugar Ray Robinson, Rocky Marciano, Willie Pep, and a host of others, all larger-than-life men moving across the snowy screen.

Casual observers who sized up young Billy as he pounded away at his improvised heavy bag were amused. "You can all laugh," he told them. "But I'm gonna be champion someday." Given how Rubin coddled his youngest son, it seemed unlikely. Rubin would order Mazie: "Make my boy something to eat!" In charge of keeping house as the oldest daughter while Dolly was at work in the fields, Mazie remembered: "We did his chores for him. He did whatever he wanted to do." Called upon to help bring in water from the pump for his bath, which was taken in the same tub that was used to launder clothes, Billy always seemed to fall asleep instead. According to Mazie, he was a "happy-go-lucky boy" who sucked his index and middle fingers until he was eleven. Whenever his behavior called for disciplinary action, Rubin intervened to spare him. Seeing Dolly take to her switch, he would plead, "Dee, leave my boy be." With a glimmer of jollity in her eyes, Mazie explained: "If he kicked the dog and the dog ran away, that would be all right. But if I kicked the dog and the dog ran away, I'd get a whuppin'."

Only years later would it dawn on Frazier how poor they had been. Kenneth Doe observed, "None of us back then were aware of that until the government told us." At the Port Royal Agricultural and Industrial School, which had been founded in 1901 by Joseph S. Shanklin of the Tuskegee Institute and later became the Beaufort County Training School, classes were given in cooking, farming, hygiene, and sewing. "We would take old clothes and turn them into quilts," said Mazie. Upon the arrival of the Christmas season, it was customary for the Fraziers to tear apart copies of the Montgomery Ward catalog or *Life* and paper the bare walls with their pages, each one a window into an inaccessible place that glowed with shiny new objects. No lavish toys were to be found under the pine that had been hauled in from outside, only bags with candy and pieces of fruit. "One year I got a black baby doll," Mazie said. "My brother got a red wagon." But whatever the Yuletide lacked in extravagances, Mazie remembered it as a "happy time," of relatives dropping by amid gales of laughter and of tables groaning with holiday fare—crab stew, fried chicken, roasted pig, topped off with sweet potato pie. Sister-in-law Miriam would prepare raccoon in later years, careful to clean and parboil it before popping it in the oven.

"Before you cook it, you've got to strip away the glands," she said. "And you've got to get one that doesn't have rabies."

And she would know that . . . how?

"Well," Miriam explained, "if it just stands there and gives you the evil eye, you better walk the other way."

So, Joe ate raccoon?

"He did when I cooked it."

Spirited roughhousing consumed Billy as a child. "He was always into stuff," said Lisa Coakley, who told the story of how one day her uncle wandered into the pigpen. "He went in there to tease the hog, thinking he could get out of the way before the hog attacked. But the hog ran underneath him and catapulted him in the air. He landed on his left arm." Holding it, he came running inside with tears in his eyes, "Momma! My arm! My arm!" But Dolly scolded, "Billy, you had no business in that hog pen!" Apparently, the arm was broken but never set in a cast, which caused it to heal with a crook in it that later enhanced his ability to throw his signature left hook. Cured of any lingering curiosity he had had of stepping back into the hog pen, Billy expelled his prepubescent energies in scraps with Annie Green, who was two years older. "You know how kids are—one thing would lead to another and we'd be wrestling in the dirt," she said. "I beat him every time! And then one day I realized, 'Hey, he can beat me.' And then we stopped." As his hours on the heavy bag piled up and his strength increased, it was not long before other boys began paying him for protection. For twenty-five cents or perhaps a sandwich, Frazer stepped between them and any "scamboogah" that happened along. (Coined by Frazier himself, the word "scamboogah" identified general lowlifes.)

School proved to be a burdensome obligation. Looking back on it, Mazie ventured Billy would have been far less bored if he had entered a tech program. From an early age, he delighted in taking objects apart and putting them back together. "My brother was a person who learned with his hands," she said. In the seventh and eighth grades at Beaufort County Training School and in ninth grade at Robert Smalls, he underperformed in his classes and scored below average in courtesy, self-control, and—ironically—perseverance. Counted absent sixty-six days during his ninth-grade year, which could be explained part by work and part by playing hooky, Frazier proved to be a perpetual headache for his teachers, one of whom in the comment section of his report card observed: "Joseph is a poor student and is sometimes very rude and ill-mannered." While Robert Smalls High School had not yet been integrated and thus did not have the resources that white schools then had, Kenneth Doe remembered it was staffed by educators "who would not let you do less than you were capable of." But Frazier did not see the point of any of it. "I was just there taking up space," he would say years later. Chances are that school would have held more appeal to him if Dolly had not prevented him from playing football for fear of injury, just as she had forbidden him from wading in ponds for fear of drowning. He did not go back for his tenth-grade year.

Whatever Joe lacked in the way of a formal education was offset by practical knowledge that came with hard living. Given the arc of his childhood experiences, it is no wonder that an unnamed friend years later told *The Saturday Evening Post*: "He was always a brute, that boy, always a brute." By his early teens, he had been tossed in the air by that hog (also booted in the head by a mule); spun a car out of control, flipped it over seven times, and emerged without a scratch; proved himself adept at beating back various scamboogahs; and become intimately acquainted with the pleasures of the opposite sex. "I never had a little boy's life," Joe would say years later. Manhood came to him at an early age, the imperatives of which were passed down by the example set him by Rubin, who—if he could bring himself to repel the invitations of other women—stood by his commitment to provide for his wife and offspring, even if it required breaking the law by running a still.

Even after Prohibition ended in 1933, it remained illegal to operate unlicensed liquor dispensaries, punishable by fines and/or jail. For the poor in the community, distilling liquor was still a way to pocket some small amount of extra income. Highly commercial entrepreneurs were called "blind tigers," who could be found traveling on the back roads of South Carolina by the light of the moon, their jugs of liquor distributed around the car in a way that would not leave the rear bumper drooping from the weight of their forbidden cargo. "A lot of corn liquor was made around here," said Rowland, the local historian. "Not at all of it was for sale, though." With the help of a government coupon that allowed him to load up on sugar—he would use five pounds of it per batch, along with ten pounds of corn and water—Rubin would let his blend sit in a fifty-five-gallon drum until it turned sour. He would then divide it into five-gallon cans and haul them to the still, where he would pour it into a boiler that sat over a fire. The still would then be capped and the distillation process would occur. "Good, strong liquor," Joe would call it years later. To keep away prying eyes on days when they were not aided by the cover of fog, Joe would wave a piece of cardboard at the rising column of smoke in order to disperse it.

The bootleg operation remained a remunerative enterprise as long as Rubin could steer clear of legal entanglements. Either Rubin would deliver the liquor (and stop in and join the lady of the house for a "drink") or customers would drop by the house to pick it up. According to Mazie, they sold a "jill" (the size of a baby-food jar) for twenty-five cents, a half-pint for fifty cents, a pint for a dollar, a quart for two dollars, and a gallon for eight dollars. But Rubin found himself at odds with law enforcement and was arrested in February 1946 for being in violation of the state liquor code. He told the judge, "Look, I have children and I'm not going to see them freeze or starve from the lack of things I know money could buy. You can throw me in jail, but I'd do it again." The judge sentenced him to four months on the public works crew. Mazie remembered he worked as a cook on the chain gang. "They worked to keep the highway clean," she said. "We'd sit across the street and watch them. And then we'd eat with them. They'd build a fire and cook a big pot of lima beans and rice."

Upset that her husband had been jailed, Dolly told him, "I think you should get out of this, try something else." Rubin replied, "What can I do? Where can I go?" But it was not just the looming presence of the sheriff that concerned Dolly. There were the all-night parties and the women who attended them. Though Mazie would call him "the kindest man you would ever want to meet," someone who would pass his own plate of food to a hungry person, she conceded: "Daddy was hell on wheels." With Joe in tow, he would go out on a Friday evening and come back hours later to find Dolly standing at the door, her eyes ablaze with fury. "Where you been?" she would holler. Only five foot six, she would begin batting Rubin across the room as he covered up. "We had to step between them," said Mazie, who added that her parents loved each other and that there was never a concern that either would leave. Dolly would shoot Billy a look and say, "And you! You are gonna end up just like him!"

Never one to hold her tongue or back down in an argument, Dolly drew a red line when it came to the expression of racial vitriol in her house. Growing up around her in the 1960s, grandson Rodney Frazier would remember once seeing some depiction of white antagonism on television and exclaiming, "Those crackers are treating us like dogs!" only to have his grandmother admonish him. "Listen," the old woman would say, "so long as you live with me, there will be none of that. God sits on his throne and looks upon his children as equal. If there is wrong in the world, just reward will come in the end." To her offspring and theirs, she preached a gospel of fairness and respect, certain in her belief that "right would prevail over wrong." But she understood that day was still far off, which is why she lectured her young on how to get by in the South: Never look at a white person in the face. When you pass them, keep your eyes down on the ground. And never, ever give a white woman even a passing glance, which was what led to the lynching of fourteen-year-old Emmett Till in Mississippi in 1955. Rodney Frazier explained, "When something happened, it heightened the fear in the family. We would say: 'Did you hear what happened to so-and-so? The same thing can happen to you.'" He shrugged and added, "So you went along to get along."

Contrary to the perception that Joe escaped Beaufort a step ahead of a lynch mob, the truth is far from cinematic and played out over a period of years in the deep shadow of Jim Crow. Indignities accumulated upon him in a slow drip, day by day. One either had to become inured to it, or pack up and leave, as Langston Hughes observed in his poem "The South": *And I, who am black / Would give her many rare gifts / But she turns her back upon me. / So now I seek the North— / The cold-faced North. . . ."* Frazier left not because he was in any imminent danger but because he sensed that by staying he would indeed come to a disagreeable end. By temperament, he did not like being pushed around, nor did he like seeing anyone else pushed around. In an interview with *Esquire* in 2006, he would point to an incident that occurred at the Bellamy Farm in which he came to the aid of a fellow black youth who had accidentally "screwed up one of the tractors" and found himself in a heated exchange with one of the foremen. Given the peril that such encounters held and knowing her boy, it became clear to Dolly that he would be better off elsewhere.

Whatever contrived harmony existed between the races hinged on the adherence by blacks to a wide range of humiliating inequities. Even if the white hierarchy in Beaufort adopted a somewhat more benevolent stance on race relations than in other jurisdictions in South Carolina, such forbearance did not come with an engraved invitation to Sunday dinner. "We stayed in our lane, and they stayed in their lane," said Tom Bolden, a Beaufortonian who married into Dolly's side of the family. Whenever there would be episodic conflict, the white establishment placed the blame squarely on President Franklin Delano Roosevelt and the New Deal. The *Beaufort Gazette* shouted in a headline in March 1944: WHO IS RESPONSIBLE FOR THE RACE TROUBLE? The *Gazette*—the very same paper that published an ad from a local seafood distributor that supported the war-bond effort by exclaiming, "Every Fish We Wrap, You Help Kill a Jap"—placed the blame on "the Dirty Politicians in the Country." An unsigned editorial argued: "We are having no trouble with Colored people except which [is] agitated by outsiders." Blame fell to "Mrs. Roosevelt," who "wants social equality in the South to help put over the Fourth Term of the New Deal." Frazier would remember being compelled to eat lunch out in the field, "like a bow-wow." And when he showed up at work each day, it was always the same exchange, which Frazier would always get a chuckle from retelling:

"Workingman arrives, says, 'Good morning, boss.'

"Boss says, 'To the mule.'

"Comes lunchtime and the workingman says, 'Lunchtime, boss.'

"Boss says, 'One o'clock.'

"Quitting time and the workingman says, 'Good night, boss.'

"Boss says, 'In the morning . . .'"

While young Billy did not go looking for trouble, he did not flee when it came looking for him. In his autobiography, he observed: "Big things, little things: Beaufort never stopped letting you know you were a nigger." Given his standing as local bruiser, street fights became commonplace. There was always some scam-boogah who stepped up to challenge him. "He did not back down from anybody," said boyhood friend Isaac Mitchell. "White or black, he'd fight if you wanted to fight. He wasn't a bully, but he didn't allow anybody to bully him." Frazier liked to tell of an encounter he had at age fourteen with an older white teenager, who shouted at him from behind the wheel of his car, "Get out of the street, nigger!" Billy shot back, "Come and do something about it, cracker!" With a big crowd looking on, Billy floored his tormentor with a left hook and continued hammering him with punches as he lay on the ground. When the young man begged for mercy, Billy let him up. The two shook hands. On another occasion, Billy whipped a marine who had taunted him with racial slurs and did it so impressively that the marine declined to press charges when the police showed up. The bloodied leatherneck told Frazier, "You beat me fair and square."

Luck would have it that in this case Frazier circumvented the judicial system, which in Dixie was a crapshoot if you were black and either the victim or alleged perpetrator of a crime. Not a great deal had changed since the late 1880s, when it was not uncommon to come across published accounts of racial disparity in the justice system: *Two white men and a black man go into the woods. The black man is shot dead. The white men testify later in court that they had heard "a noise" behind them that they feared was "a snake" and fired upon it. They are acquitted.* Thus, it would come as no surprise when fourteen-year-old George Junius Stinney Jr. was arrested in March 1944 on circumstantial evidence in the slaying of two white girls, ages eleven and eight, in the rural South Carolina community of Alcolu. With only an unrecorded and unsigned verbal "confession" coerced from him with an ice cream secured for him by a deputy, Stinney was convicted in a one-day trial by an all-white jury that took just ten minutes to deliberate. No appeal was pursued by his white public defender. The ninety-pound youth was so small that they had to place a Bible under him in order to hook him up to the electric chair. He became the youngest person ever executed in the United States. A judge would overturn the verdict in 2014 on the grounds that Stinney had been denied a fair trial, provided no effective defense, and deprived of his Sixth Amendment rights. Observed Kenneth Doe, who became a pastor: "This was a heartless land."

Word of what happened up in Alcolu only gave a face to the fear that was in the wind, of how disposable black lives were in the hands of judges and juries. The very same spring that Stinney was being rushed to judgment, Edward Feltwell, the sixteen-year-old white son of a marine warrant officer at Parris Island, confessed to the rape and murder of an eight-year-old white girl. Two weeks before Stinney was executed at the Central Correctional Institution in Columbia, Feltwell received a twenty-year prison sentence. Exactly how much of that he served remains unclear. The Federal Bureau of Prisons cannot verify that he completed any of it. Records indicate that he was married in North Carolina in 1957, worked as an electrician, and died in Jonesville, in 1995. He had two daughters. None of this came as any surprise to Doe, who observed that the scales of justice were only balanced if the black community could find a way to apply leverage. To elucidate, Doe cited an incident that occurred in the 1920s. "A white boy spit on a black boy," he said. "The black boy grabbed a stick and broke his arm. Of course, the law came. But this is what happened: the adult men of the black community went to a white grocer in the area and said, 'If this boy goes to trial, your business is closing down.' And he never went to trial."

Given how at odds his disposition was with the prevailing culture, it is no surprise that Frazier would weigh his circumstances and conclude that "there was nothin' but bad times ahead." Moreover— and perhaps more significantly—there was no foreseeable way that he could better himself in Beaufort. Employment was menial, spo-

radic, and low-paying. Upon running afoul of his boss at the Bellamy Farm, he found work unloading crates at the Coca-Cola plant
and later got a job raising rafters at a government housing site on
Parris Island. To pick up extra money, Frazier and a friend, B. A.
Johnson, seized upon broken-down cars that had been abandoned
by marines on the roadside until payday. The two would siphon off
the gas for their own use and strip the vehicle of its parts, which
they sold. With the exception of his disorderly behavior in street
fights, this juvenile dalliance with larceny was as far as Frazier would
step toward the wrong side of the law. No indication exists in South
Carolina records that he was ever arrested.

Joe looked upon himself as a man at that young age. With what
he called "the mind of a twenty-two-year-old" at the age of just
fourteen, he found himself in relationships with two teenage girls,
Rosetta Green and Florence Smith. The young women became rivals for his attention, if not a commitment. With Rosetta looking
on, Florence asked him: "Billy, who do you love?" Frazier told her
he loved them both. According to his autobiography, he believed
that "a man can love as many [women] as he can love." Neither
woman wanted any part of such an unconventional arrangement,
yet each would bear him not one child but two before he turned
nineteen. Although his entanglement with the two women was re-
solved when Frazier eventually chose to marry Florence, it would
remain a complicated arrangement among the three.

Florence Smith was an attractive young woman with cute
braids. By nature, she tended to be timid. She lived fifteen miles
away from Laurel Bay on St. Helena Island, the only daughter of
Hector and Elise Smith. Hector worked as an electrician on Parris Island, Elise as a housekeeper. Along with Florence—who they
referred to endearingly as "Lady"—they had three sons, including
one who drowned as a young child. Hector kept a close and loving
eye on Florence, whom he hoped to see one day attend college, as

she had been at the top of her high school class and had the ability to go far. So when she turned up pregnant, he became irate, eyeing Billy with scorn and sizing him up as a hooligan from Laurel Bay.

Even Frazier would have had to concede that he was something less than optimal son-in-law material. With no prospects of a steady job, two children on the way by two women, and a disposition that placed him in the crosshairs of conflict, he began looking beyond the county line for a place where he could settle down and better himself. His brothers Bubba and Bozo had gone to Central Florida to work in the orange groves; Bozo later hired on as a longshoreman in Tampa. Mazie and Bec had traveled north to Philadelphia, where both found jobs in the garment industry; Bec later became a union organizer. Tommy settled in New York, where he also worked in the garment industry. Upon dropping out of school, Frazier arranged to join Tommy and his wife, Ollie, in New York. He asked his friend Isaac Mitchell to come along.

"Oh, man, you can do better going north," Billy told Mitchell.

"Yeah, but I'm still in school," Mitchell replied. "You know my mother and father ain't gonna let me, especially my father. I can't go home and say I'm packing up and leaving."

Billy shrugged and said, "I gotta go. I gotta help my family."

At the beginning of the Great Migration, the journey from the Sea Islands to New York was engaged by steamboat, by way of Savannah or Charleston. ("There's a boat dat's leavin' soon for New York, come wid me, dat's where we belong, sister," was how Ira Gershwin and DuBose Heyward enshrined the passage in the lyrics of *Porgy and Bess*.) By the 1920s and '30s, the preferred way of traveling was by train on "The Chickenbone Special," so-called because black passengers who could not get served at railroad restaurants would carry shoeboxes full of food; all that remained by the end of the trip were some chicken bones. By the 1940s and '50s, the Greyhound bus became a popular alternative for passage. Choosing to go by bus, fifteen-year-old Billy Frazier borrowed the fare from his cousin Charles Middleton. With a single suitcase that contained a few odds and ends of apparel and a change of church clothes, he boarded the Dog by himself in Beaufort for the journey to New York and sat in one of the middle rows, only to be ordered by the driver to move back as additional white passengers got on at each stop. As the bus coughed and wheezed up Route 301 and the hours crawled by, he peered out his window at the passing shopfronts and signage, the small towns giving way to vast fields and then to ever-larger cities. In his lap was a paper bag with some cold pieces of fried chicken.

CHAPTER TWO

THE HAMMER OR THE NAIL?

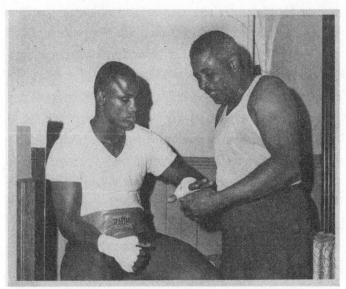

Joe and Yank Durham, 1966. *Philadelphia Bulletin*

F rom across the crowded and noisy gym at the Twenty-Third Police Athletic League in North Philadelphia, his voice boomed with an authority unlike any other. No one had the set of vocal cords that Yancey "Yank" Durham possessed, or used them with such agility as a tool of persuasion. He had a basso profundo that oozed honey. Assiduously, he polished his delivery by taking correspondence courses to enhance his diction, which was always precise in polite conversation yet invariably peppered with profanity. *New York Times* columnist Red Smith once observed: "If a person who did not know Yancey Durham heard him talking in the next room, he would assume that the voice belonged to an actor or preacher or con man or politician. . . . Yank was a little of each." With a slender mustache, his hair spotted with flecks of gray, he dressed in tailored suits, drove a Cadillac, and favored big cigars. He carried himself with the imperial air of a U.S. senator.

With a towel over a shoulder and a Q-tip behind an ear, Durham had labored in relative obscurity for years, his eyes always on the gym door in search of a young fighter who could turn a dollar. Raised across the Delaware River amid the urban decay of Camden, New Jersey, he had had nine bouts as an amateur fighter, not including eleven more unsanctioned events on what he called "bootleg cards." On one he shared billing with "Two Ton" Tony Galento, who had been flattened by Joe Louis during his "Bum of the Month Club" tour but not before he had dropped Joe himself. Durham earned ten dollars for these appearances. "They'd give you some flunky name and you'd fight," he said. World War II interrupted his fledgling career and effectively ended it when a jeep ran over him during an air raid in England. With both legs shattered, his skull fractured, and assorted ribs broken, he was in and out of hospitals for two years and would never overcome a slight limp. He found employment as a welder for the Pennsylvania Railroad, which would later provide him with a pension that supplemented the disability pay he received as a veteran. Through the 1950s and early 1960s, he was a fixture at the Twenty-Third PAL, where he worked with assorted amateurs with the objective of turning them pro. None of them would come within even hailing distance of the brass ring.

Philadelphia gyms were abattoirs in which only the bravest survived. They were scattered across the city in run-down buildings, in walk-ups above auto shops or in basements, and were known for the unpoliced mayhem that occurred in those small sixteen-by-sixteen-foot rings during sparring sessions. Crowds would stream in off the street as word spread of an impending encounter, and bets would be placed on the outcome. Veteran trainer George James told me, "When Georgie Benton was boxing Harold Johnson, you could hear a pin drop. Both of them staring each other down, daring one another to throw a punch so the other could counter off it." In 1960, outweighed by sixty pounds, Benton found himself in a sparring session with the formidable Sonny Liston, two years before Liston pummeled Floyd Patterson for the world heavyweight championship. "Sonny was a brute," Benton told the *Philadelphia Inquirer.* "We got into the ring and he started swinging at me. I had to outbox him, sticking and moving, sticking and moving." By the third round, Liston was so enraged that he had not yet pinned down the elusive middleweight that he began launching wild shots at his head. To the eternal relief of the understandably petrified Benton, his manager stepped in and stopped the action before Georgie ended up a casualty of the so-called Philadelphia "gym wars." Even with headgear and oversized gloves, the savage engagements in these sparring sessions could be more perilous than the actual paid events for which the boxers were preparing. By whipping an opponent in these informal settings, it conferred upon a fighter a certain street cred—"King of the Gym." And yet doing so did not herald longevity for either combatant, as *Philadelphia Daily News* columnist Stan Hochman once observed: "Philadelphia gym wars shortened more careers than cocaine."

The Twenty-Third PAL stood at street level at Twenty-Second Street and Columbia Avenue, with big windows looking onto two rings in a long, narrow interior. Overseen by Hammond E. "Duke" Dugent, a genial cop who looked upon it as a sanctuary from the corrosive elements of urban life, the gym produced some of the top headliners of the 1950s and '60s, including Gil Turner, Sugar Hart, Bennie Briscoe, and Gypsy Joe Harris. The star-crossed Gypsy Joe was just a boy when he hid in the doorway with a pursuer on his heels, only to glance through the window, grow curious, and go inside. Yank would have a hand in training him, along with

Briscoe and others in the 1960s. But only when Durham became acquainted with Joe Frazier did he come face-to-face with his destiny. When Dugent pointed him out in early 1962, Durham was not particularly impressed. Joe was flabby and pigeon-toed. He had come to the gym to drop twenty or thirty pounds, so he could get back into some clothes he had grown out of and perhaps enhance his appeal to women. But appearances were immediately forgotten as soon as Durham and his partner Willie Reddish eyed him hitting the heavy bag. The sound that emanated from the blows Joe tagged it with produced the same singular acoustic effect that Ted Williams created when he squared off on a baseball.

"Joe was slapping at it like a girl," said James, who was at the Twenty-Third PAL that day. "But when he hit it with that left hook—BOOM—you could hear it across the street. Yank and Willie just looked at each other and said, "Wow.""

Even though his previously transplanted relatives had welcomed Billy with open arms when he stepped off the Dog two years before in New York, the "cold-faced North" that Langston Hughes had warned of did not. He moved in with his brother Tommy, sister-in-law Ollie, and their two children in their three-bedroom apartment in Manhattan. Tommy had gone to New York a few years before and found work in the garment district. Billy's cousin Ginger Bolden escorted him to the Employment Agency Center at 80 Warren Street, a five-story building on the southern end of Tribeca that housed dozens of low-end employment agencies that were the first point of contact for many migrants up from the South. In his 1962 book *The Other America*, author Michael Harrington referred to it as a "slave market" that provided "the work force for the economic underworld of the big city." Of the abject crowd of humanity that congregated there, which included alcoholics and the mentally challenged, Harrington observed: "Most of the people at 80 Warren Street were born poor. They were incompetent as far as American society is concerned, lacking the education and the skills to get decent work."

Work was hard to come by and even harder to keep. Unable to hook up at the Employment Agency Center the day Ginger took him there, Billy roamed the city and found odd jobs, including one in a sheet metal factory. "He had little jobs—none of them for too long," Ollie Frazier told me. To get by, Billy borrowed money from Tommy and dipped into his old Beaufort playbook by stealing idle cars parked along the street with his friend B. A. Johnson, who was bedding down with relatives in Bedford-Stuyvesant. In his autobiography, Frazier recounted how he and Johnson discovered a junkyard in Far Rockaway that paid them fifty dollars a car without asking any questions. They averaged one heist a week. "Late-model beauty or old bomb, them junkyard boys didn't care. 'Here's fifty bucks, see ya.'" On the weekends, there would be house parties across the city that distracted him from the aching displacement he endured. "He could never get settled in New York," said Ginger. Sensing that his brother was perilously adrift, Tommy suggested that he move to Philadelphia, where Mazie, Bec, and others had settled. In looking back on it years later, Mazie said, "Tommy knew that if Billy stayed in New York any longer he would end up in trouble." His aunt Evelyn Peeples agreed to give him a room in her three-story house in North Philadelphia.

Beyond the problems he had squeezing into his clothes and, as he would explain later to *Playboy*, the even more pressing concern of not having his "way" with women, he was encouraged by Mazie and her husband, James Rhodan, to utilize the Twenty-Third PAL as a safe haven. Both were concerned Frazier would fall under the spell of his paternal uncle Oliver. Known fondly within the Frazier clan as "Uncle Cadillac"—he always seemed to be behind the wheel of a new one—Oliver was a tall, sharply dressed street hustler whose eyes were always hidden behind dark shades. "Unc never seemed to have a job but always had money," said Lisa Coakley. His chief pleasures were playing cards and running with women, one of whom owned a speakeasy in North Philadelphia. "Billy was just like him," said Mazie. "They were fancy free." Only too aware of how easy it would be for him to end up with the wrong crowd—and how devastating that could be, given that drugs were taking hold in the city—Mazie sat down with Billy to express her concerns.

"Look, if you get in trouble, I don't know anyone who could help you out," she told him. "I wouldn't know where to go or what to do. But I do know that the way you are going is not a good way. You have your children to think of, not just yourself. So, you have to do something to better yourself."

Billy nodded and replied, "Okay, I want to do better." Mazie would remember, "He was just like a little boy again."

James Rhodan took an enthusiastic interest in Billy. "They fell in love with one another," Mazie said. "They were so close you would think they were brothers." When James came home from his job as a welder at the shipyard, the two would go jogging in Fairmount Park. And while James was in no way an aficionado of boxing, he accompanied his brother-in-law to the Twenty-Third PAL, which both he and Mazie hoped would help him form a protective alliance with the police force—just in case.

Dugent would remember the day he showed up, in an unmatched suit coat and trousers with an open-collar shirt, just prior to Christmas 1961: "It was late afternoon. He said he had just come from New York and he was out of work. He weighed 240 pounds. Most of it was in his rump and thighs. He said he wanted to fight." Never one to discourage the dream of an underprivileged young man in search of a break, even if it seemed to his eye an improbable undertaking, Dugent escorted Frazier into his office and handed him a membership form. A relative who owned a North Philadelphia barbershop gave Joe some money to buy some boxing shoes, hand wraps, and a gym bag. With the arrival of 1962, he began working out.

Frazier found a job as a butcher at Cross Brothers Meat Packers, a kosher slaughterhouse where once an eleven-hundred-pound Lancaster-bred steer escaped from a boxcar and led a rollicking parade of fifty children through the North Philadelphia streets before a plant supervisor cornered it in a schoolyard and plugged it with ten rifle shots. In a safety helmet and apron, his biceps jutting from beneath his sleeveless uniform, Frazier spent long days in the refrigerated chill handling unwieldy pieces of animal flesh and hosing pools of blood off the slippery floor. "I never minded seeing [the animals] cut," Frazier would say. "Some of 'em looked like they wanted to keep living, though. They'd try to run out of the room, throats cut and everything. It didn't bother me none." Even if the pay was poor—twenty-five dollars a day—it was dependable employment that steered him away from any temptation to stray into illegal activities. "It made him more of a man," said Mazie. "He could live like a person should live—buy food, clothes, even a car. With some money in his pocket, he could stick his chest out." But the job was not without hazards, particularly for someone who used his hands for boxing. The blades he used were so sharp, it seemed as if they could cut through tempered steel. When he sliced off the tip of the pinkie on his left hand, the wound required a skin graft and nine stitches to repair.

Florence showed up unannounced in Philadelphia in the spring of 1962. By then, she and Frazier had had their second child, daughter Jacquelyn. "I remember it was a rainy day," said Frances Morrall, who was then rooming with Aunt Evelyn. "She pulled up in a taxi and rang the bell." Florence had passed up an opportunity to attend college two years before. She had been scheduled to begin classes on the very day their first child, Marvis, was born—September 12, 1960. The acceptance letter she received would remain among her belongings as a reminder of the alternate path that had once called to her. But she loved Billy, saw the good heart in him, and remained devoted even as he wavered in his commitment to her. Aunt Evelyn invited Florence to stay with her, but only on the condition that she and Billy occupy separate bedrooms. She was not one to allow casual sexual activity under her roof. But Aunt Evelyn was away the better part of the week working as a domestic and could only police their behavior so far. When Florence became pregnant with Weatta, she and Billy were married, on June 25, 1963, at the home of a local minister, with only Frances and her boyfriend in attendance. Billy was so strapped for cash that Mazie gave him her own wedding ring to use in the ceremony.

Whatever wayward impulses he had yielded to during his young life, Frazier found an overriding purpose in boxing and applied himself to it with extraordinary discipline. To accommodate his work schedule at Cross Brothers, he would get up before dawn to run in Fairmount Park and end his day with a two-hour session at the gym. With him always during this period was James, who learned how to wrap his hands and carried his bucket and water. To help him get his weight off, Dugent would remember, "We put him on a diet. Nothing severe. Just green vegetables and steaks and stuff like that. And cut out the sweets." Along with countless push-ups and sit-ups, Dugent would get him to stand on his toes for long periods in order to develop his balance. Dugent schooled him in the proper way to throw combinations and found that Frazier had an aptitude for instruction. "At first he was bad, the next day he was great," Dugent said. While Dugent appreciated Frazier's exceptional "short power," which "came down from the shoulders through the chest," he also thought the fighter's physical shortcomings would always outweigh whatever intangibles he possessed. For a heavyweight, he was simply too small, with arms that were too short and legs that were too heavy. Moreover, he was initially a southpaw, which was frowned upon during that era; he would not convert to an orthodox stance until late in his amateur career. But when Durham saw him slam the heavy bag, he was just intrigued enough to ask: "You think we can work together?"

Though Durham drove his fighters hard, he only did so with those he was certain had the potential to grow and thrive. Caution guided him in his assessment of talent. In a column in the *Philadelphia Bulletin* years later, Claude Lewis shared a conversation he had had with a former would-be fighter named Billy Johnson, who came to Durham with the same ambitions that Joe had but none of his raw skills. "I worked my tail off and I went to Yank and I asked him to take a look at me in the ring," Johnson said. Durham took a look at him, but as Johnson remembered, "It wasn't a long look." Durham walked over to Johnson with his head down, placed his arm over his shoulder and said in a confiding voice, "I wouldn't want you to get hurt." Johnson quit boxing on the spot and became a presser in a tailor shop, where he remained safely and happily employed. Durham understood that the ring was no place to fool around, that it could be unsparing in the price that it exacted. For this attention to such ethical concerns, it was once said that Durham stood out in the unprincipled realm of boxing "like a healthy thumb on a leprous hand."

Beyond the raw power he spotted in Frazier, Durham ascertained that there was a big engine inside him. Given the here-today, gone-tomorrow level of commitment of the young aspirants to whom he had been accustomed, Durham sized Frazier up as just "another fat kid" who would disappear as quickly as he arrived, even if he did have thunder in that left hook. But it became clear as Frazier shed pounds that he had his eye on the eagle and attended to the drudgery of training with uncommon focus. On those days when Durham was traveling and Frazier had to work late at Cross Brothers, Frazier would use the key to the Twenty-Third PAL that Dugent had entrusted him with and work out to the accompaniment of forty-fives on a portable record player. Each song was more or less three minutes, the length of a round of boxing. As cars swished by on Columbia Avenue outside the gym window, he would stab his gloved hands into the heavy bag and give a grunt, as his sweat dripped to the floor and the walls echoed with songs of Cupid and chain gangs and twistin' the night away.

————

At the 1964 New York World's Fair in Flushing Meadows, New York, one of the more popular attractions was Walt Disney's "It's a Small World," a cheerful singsong display of three hundred animatronic dolls arrayed in costumes from around the world. But across the way at the outdoor Singer Bowl that May, there was nothing small about the anatomical specimen who stepped into the ring to face Joe Frazier in the Olympic boxing trials. With layers of flab spilling over the waistband of his hiked-up trunks, the six-foot-three-inch Buster Mathis wrecked the scales at 296 pounds, trimmed down from the 340 pounds he had lugged around prior to the commencement of training that spring. Boxing out of Grand Rapids, Michigan, he was light on his feet for a man his size and had impressive hand speed. Going into the three-day tournament, he was projected as the odds-on favorite to go to Tokyo and bring back a gold medal, if only because it appeared that no opponent he was likely to encounter here or abroad would be able to offset his colossal size. Observed *Sports Illustrated* writer Robert H. Boyle in his piece "At the Fair with Fat Buster": "Sitting in his corner, he looks like a melting chocolate sundae."

By 1964, Frazier had evolved into a 196-pound heavyweight with promise. Early on, as Durham would remember it, Frazier "thought he was a boxer" and acquitted himself as such. Lightweight "Classy" Al Massey worked out with him at the Twenty-Third PAL and told me he once gave Frazier some boxing tips while Durham was away, only to be admonished the following day by Dugent. Massey remembered, "Yank saw what I had done and asked Joe, 'Who showed you that?' Frazier told him. So Yank approached Duke, and next thing I hear is Duke: 'Massey!' I go over and Duke tells me, 'Joe is a puncher. NOT a boxer.'" Do it again, Dugent warned him, and he would be tossed out of the gym. To counterbalance Joe's physical shortcomings, Durham was of the belief Frazier only had one way to go and that was "straight ahead" into the chest of his opponent—chin down, eyes up, always on the attack. PAL trainer George James explained that Durham had envisioned Frazier as "a fullback." James said, "The thinking was, Joe would charge in there, like a fullback hitting the line, and never give his opponent a chance to set his foot to throw a punch." Frazier was not immediately receptive to the strategy of walking into punches. Durham would say, "He thought I was a damn fool."

Of the exceptional crop of amateurs who passed through the Twenty-Third PAL in the early 1960s, Dugent used to say that while no one was as dedicated as Frazier, or was more a "killer" than Bennie Briscoe, none surpassed Gypsy Joe Harris when it came to sheer talent. From his clean-shaven skull down to the tassels on his boxing shoes, five-foot-six Joseph Louis Harris was a virtuoso who never showed you the same move twice in the ring. Away from it, he was a free spirit who roamed the nightscape in a white cowboy hat. At the gym—that is, when he was there, and not in a pool hall or a bar—he would give away fifty pounds and spar with Frazier, who early on had flattened the number-two and then the number-one heavyweights in training there. Frazier remembered in his autobiography that he himself "handed out asswhuppings like lollipops" in his sparring sessions. But Gypsy Joe would back away from no man.

"Go get him, Gyp!" someone shouted from ringside. Massey remembered that Gypsy had Frazier trapped in the corner. When Gypsy stepped back to load up his punches with leverage, Frazier walloped him with a left hook that drove Gypsy clear across the ring into the turnbuckle. But Gypsy did not go down. Amid the exhortations of the onlookers at ringside, Gypsy stared daggers across the ring at Frazier and sneered, "Okay, you motherfucker, you want to fight?" Upon hearing that, Dugent dashed from his office, threw his arms around Gypsy Joe, and escorted him to safety.

Joe loved Gypsy. According to Denise Menz, he saw himself in Gypsy, perhaps that part of him that bridled in the presence of hard boundaries. One of four children who had been abandoned at an early age by their father, Gypsy was working as a delivery boy for a grocery store in North Philadelphia when he found his way to the Twenty-Third PAL. As he hustled out of the store one day toting a bag, he bumped into a man with an ice cream cone in his hand. When the cone splattered on the pavement, the man chased him down the street and into the gym. Gypsy later told Robert Seltzer of the *Philadelphia Inquirer*, "I ran like Jesse Owens." For years, no one at the Twenty-Third PAL would know that he was blind in one eye. Only years later would he explain how it had happened: On Halloween in 1957, he had snatched a bag of candy from a friend and took off. When Gypsy turned and looked back, the friend heaved a brick that struck him in his right eye. Ten days in the hospital followed, but the eye could not be saved. He was eleven years old.

Gypsy Joe became skilled at covering up his impaired vision. In the ring, he compensated for it by concocting an array of daring moves that saw him dip, bend, and twist in order to keep his good eye fixed on his opponent. George James said, "It was amazing how he could protect himself and still fight with just that one eye." By his own admission a coward out on the streets, where he armed himself with a four-and-a-half-inch switchblade that he called his sword, Gypsy was fearless once his gloves were laced up and he bounced through the ropes. While he possessed none of Frazier's discipline, he shared the same goal of a bigger and better life. Even as he would remain anchored in the unforgiving ways of the street, where he gorged himself on hoagies by day and shadowed the perfumed scent of the "foxes" until dawn, he saw a gilded future stretched out before him, one that included a wardrobe big enough that he could "change clothes twenty times a day." He once said of boxing: "You can be the hammer, or the nail. I wanted to be the hammer." Sadly, it would be his destiny to be the nail.

Only one of the two Joes would end up competing for a berth on the 1964 U.S. Olympic boxing team, and it would not be Gypsy. Though he had won the Middle Atlantic Golden Gloves the two previous years, he was beaten in the eastern regional Olympic trials in what was criticized by some observers as an unfair decision. But Frazier pushed himself harder than ever during his run for Tokyo. With the 1962, 1963, and 1964 Middle Atlantic Golden Gloves Heavyweight Championships behind him, he doubled his roadwork from three to six miles each morning. He wore special boots that weighed six and a half pounds each and carried eight-pound weights. He jabbed and whipped his hands out in a flurry of combinations as he jogged the sloping terrain of Fairmount Park. The extra effort led to an increase in stamina, which would enable him to pour on the pressure as he stalked his opponents. Moreover, his punching power seemed better organized. At the eastern regional Olympic trials, he stopped four opponents on back-to-back days. At the Olympic trials in Flushing, he added his fifth straight knockout by dumping the 1963 national Amateur Athletic Union (AAU) heavyweight champion Wyce Westbrook out of the ring in the second round with a left hook to the jaw, followed by a left/right combination; Westbrook required fifteen stitches in his mouth. In the semifinals the following day, Frazier floored Clay Hodges with a chopping right and won by technical knockout in the second round. That placed him in the finals against Mathis.

Columnist Jack McKinney lampooned Buster Mathis in the *Philadelphia Daily News* as "too big to be a man, too small to be a horse." They called him "Little Big Daddy," a salute to NFL defensive lineman and fellow Detroiter Eugene "Big Daddy" Lipscomb. Born the youngest of eight children in 1943 in Sledge, Mississippi, Mathis grew up in the Motor City and found himself orphaned at age fifteen, at which point he became the ward of sign painter Paul Collins and his wife in Grand Rapids. Along with working his way up on the amateur circuit as a boxer, he played some defensive tackle for the semipro Grand Rapids Blazers. By all accounts, he was an appealing young man—a jolly sort—who blamed his weight gain on his weakness for soul food, particularly fried chicken and pinto beans. As the presumptive Olympic heavyweight gold medalist, he said he dreamed of one day tangling with Cassius Clay, of whom he said: "I really believe I excel him at everything except purtiness." Frazier liked Buster, enjoyed his company, yet dismissed his size as "baby fat." Frazier had scouted him in the semifinals and concluded that he had "nothing on his jab" and that his hook was so long that it could stop and take on passengers. Frazier said, "My only problem will be getting out of the way when he falls."

Hours before they were scheduled to step into the ring, it appeared that Mathis would be forced to withdraw because of high blood pressure. When it was checked at 3 P.M., his systolic pressure was an alarming 195, apparently due to his excessive weight. But he lay down for three hours and it dropped to an acceptable enough level for the bout to proceed. While Frazier had said that he expected it to be the easiest of the tournament, he found himself flummoxed by the sheer bulk of Buster, who outweighed him by a hundred pounds. Chopping him down would be a project on the order of angling a piano up a narrow stairway. Referee Roland Schwartz told then–*New York Herald Tribune* columnist Red Smith: "Joe feinted a hook to the body and threw a hook to the chin and I thought, 'Oh-oh. Get ready to start counting.' But Buster never blinked." Whenever Frazier penetrated his perimeter, he encountered an undulating cushion of flab. With Frazier on the attack, Mathis counterpunched effectively and revealed surprising agility, which prompted ringside announcer Don Dunphy to observe: "He moves almost like a middleweight." Frazier continued to apply pressure in the second round, but was penalized two points by Schwartz for low blows; Joe would later complain that Buster wore his trunks almost up to his nipples. Up on his toes in the third round, Mathis wobbled Frazier with a wild right to the head. Had it been a longer bout, Frazier very well could have worn down Mathis over a period of rounds. But Buster was awarded the decision. Then–*New York Journal American* reporter Dave Anderson told me: "Somehow the judges voted for Buster Mathis. They seemed to be dazzled by his ballet, instead of the punches. I wondered if we were watching the same fight."

Frazier vowed he was done with boxing. "Pitty-pat, pitty-pat. He hit me with his best shot and it was nothing," Frazier told flyweight Bobby Carmody, who would capture the bronze medal in Tokyo and three years later die in an ambush while serving with the U.S. Seventeenth Cavalry Regiment in Vietnam. Gene Kilroy, who would later oversee business affairs for Ali, had joined his old army buddy Carmody at the trials that week in New York and remembered how dejected Frazier was. "He said he was going to go back to Philadelphia and just get a job," said Kilroy. Carmody reminded Frazier that he had the ability to be a good pro. Frazier just shrugged. But as he packed his gear and prepared to head to the team bus that would take him back to the Hotel Commodore, the beaten and forlorn young man did something that impressed Anderson, who was gathering notes in the dressing room. Frazier spotted *Philadelphia Bulletin* boxing writer Jack Fried, who had been attentive to his amateur career. Joe shook hands with him and said, "Thanks, Jack, for all you did for me." Anderson remembered thinking, "This is a guy with some heart."

———

At the epicenter of North Philadelphia on a humid Friday evening in August 1964, a quarrel erupted between Rush Bradford and his wife, Odessa, in the front seat of their car. Odessa grabbed the steering wheel and held her foot on the brake, which caused the vehicle to come to a stop in the shadow of the Twenty-Third PAL, at the intersection of Twenty-Second Street and Columbia Avenue—also known as "Jump Street" or "The Ave." With traffic snarled amid a cacophony of horns, a city cop leaned into the car to extricate the woman from it, only to be seized upon from behind by a passerby. Mayhem ensued. Bricks and bottles flew through the air as police backup converged on the scene. Rumors circulated that the cops had beaten to death a pregnant black woman. Over a seventy-two-hour period, across twenty blocks of North Philadelphia, looters ransacked storefronts and incinerated buildings in an event that left one person dead, 339 injured (including one hundred police), and 775 in custody. Photographed in a half-shell riot helmet in the *Philadelphia Daily News* was Deputy Police Commissioner Frank Rizzo, who would use the Columbia Avenue riots as one of the underpinnings for his brass-knuckle climb to power.

The Twenty-Third PAL was the only building in the immediate vicinity that did not have its windows shattered by rioters. Even incorrigible hoodlums appreciated the good that "Mr. Duke" performed in the community. In a *Sports Illustrated* cover piece written by my father that appeared on June 19, 1967, Gypsy would remember the looting and admit that he had a small hand in it himself. "I finally couldn't help it when I see these three men tryin' to load a refrigerator in a car," he said. "The cops came and hauled them away, and while they were bein' taken this chick jumps up on top of the refrigerator and starts screamin', 'Black men unite!' I said to myself, 'Baby, I'm gonna unite all right,' and then Gypsy start diggin' in himself." Atop the refrigerator, the woman began shouting down local NAACP leader Cecil B. Moore, who appeared on the scene to quell the rioters. That his words were not heeded would lend weight to the belief by Rizzo and others that the civil disobedience had been the work of subversive "black militants."

Frazier held a pragmatic view of the rampaging that occurred in Philadelphia, New York, and elsewhere during the summer of 1964: to Frazier—who kept "a low profile" during the riots, according to Mazie—it was hard to understand why someone would burn down their own neighborhood, the very place where they lived and worked. Notwithstanding, he could appreciate the rage that racism triggered, how it incited a passion to push back. While Frazier was not one to back down from any challenge to his pride as a man, he was also not one to court open confrontation with white authority. Unless he was provoked, he held a conciliatory if wary view of the white establishment, knowing that it stood squarely between the poverty from which he sprang and the actualization of a better life. The only bombs Joe would ever throw would be in the ring.

Whatever impulse Joe had to quit boxing in the wake of his defeat to Mathis dissolved with the intervention of Durham and Dugent, both of whom impressed upon Frazier that there were better days ahead. Invited to join the Olympic squad as an alternate, Frazier was initially uncertain if he could afford it, given that leaving home to join the team would jeopardize his job at Cross Brothers. The one-hundred-dollar-a-week position that Florence had as a checkout clerk at Sears was far from enough for her and the children to get by on. Dugent elicited the aid of Rizzo, who in turn called an owner of Cross Brothers and was assured that Frazier would have a job when he came back. Thus, Frazier dipped into the few dollars he had saved to pay for his passport and inoculations, boarded a plane to California the first week of September, and reported for training at Hamilton Air Force Base, near San Francisco. There, he became reacquainted with Mathis, who was as big as ever, playfully unfocused, and in no apparent mood to seize upon an opportunity that Frazier himself still craved. By that point, Frazier had but two outside chances of competing in Tokyo: Buster would have to suffer a disabling injury or Frazier would have to drop almost twenty pounds to box as a light heavyweight.

Only twenty-three days before the Tokyo Games commenced,

Frazier and Mathis fought again in an exhibition at Hamilton Air Force Base. Again, Frazier lost, this time by a split decision. The San Rafael *Daily Independent Journal* observed in its report the following day that Buster "shook like a jar of jelly" and had problems keeping up his trunks. The packed gym roared in laughter as he stopped and yanked them back in place. As he had in May, he exhibited impressive footwork and counterpunched effectively across three rounds. According to the *Daily Independent Journal*, Frazier hit him with "tremendous jolts." None of them had any effect. Frazier would later tell *Washington Post* columnist Shirley Povich, "My gloves sank in his belly up to my wrists and nothing happened." But Mathis broke a knuckle on his right hand. Two days later, it was announced that his hand would be in a cast until a day before the opening ceremonies, and he would be replaced on the team by Frazier. Gene Kilroy heard from Carmody that Mathis would be sidelined and called Frazier with the news. Frazier exclaimed: "Oh, God! Thank you, God!"

Japan looked upon the Tokyo Games as far from just an extravagant sporting event. Nineteen years had passed since the United States secured its surrender by dropping nuclear weapons on Hiroshima and Nagasaki, and it viewed the Olympic Games as an occasion to position itself with renewed dignity upon the world stage. To avoid any possibility of bringing shame upon itself, Emperor Hirohito issued initiatives to clean up the city, which included a sweep of the local sex trade. In his article on the Games in *Sports Illustrated*, Jack Olsen quoted an unidentified athlete who said he had gone to a Turkish bath in the Ginza and complained: "They gave me a Turkish bath!" Even known pickpockets were forewarned that their pictures had been circulated and that close tabs would be kept on them. But the Japanese government did not anticipate the lawlessness that would be engaged in by the athletes themselves, who plundered Tokyo stores on shoplifting sprees that challenged the congeniality of their otherwise agreeable hosts.

Friendly young American women could be found at the International Club in the Olympic Village. Frazier told Olsen, "They do The Crossfire, The Pony, The Hunt, [and] The Monkey." Then, he added by way of explanation: "They're all dances, you know?" In between his obligations with the team, he hung out with Carmody, who in long-distance phone calls with Kilroy said, "This Joe Frazier, what a good guy. We're having a lot of fun!" Carmody told Kilroy that when the team members washed their clothes together, Frazier would collect any unused soap and send it back to Florence in Philadelphia. Florence received letters from her husband each day—sometimes two a day. In them Frazier replayed the events of his day in Tokyo, said he loved her, and reminded her, "Keep the children off the streets." Insofar as the opponents he would face were concerned, he told her not to worry, writing: "They fight like girls." Cockiness oozed from him. When a teammate asked, "You gonna get the gold, Joe?" he would point to his chest and reply, "You're lookin' at it, boy."

Four opponents stood between Frazier and the gold medal. Far from the smooth sailing he had anticipated, there were some challenges as the tournament unfolded. With the exception of a relatively easy first-round victory over Ugandan George Oywello—whom he stunned with left hooks and stopped on a technical knockout at 1:35 of the first round—Frazier found himself in fleeting jeopardy in his quarterfinal and semifinal bouts, against Australian Athol McQueen and Russian Vadim Yemelyanov, respectively. McQueen nailed Frazier with a right hand to the chin in the first round that dropped him to his haunches. Frazier had never been knocked down before, which became a point of pride for McQueen. "Geez, he was a tough bugger," the Aussie said years later. Frazier answered with a barrage of body blows and stopped him on a technical knockout forty seconds into the third round. Yemelyanov wobbled Frazier in the first round of the semifinals with a solid right to the chin, but Frazier floored the 213-pound

Leningrad soldier with a left hook in the second round and battered him from corner to corner until the referee halted the fight at 1:59 of the round. Frazier would be the only one of ten American boxers to advance to the finals.

Awaiting Frazier there would be German Hans Huber, a thirty-year-old bus driver who had been rejected by the Olympic wrestling team. Unbeknownst to anyone, Frazier had injured his left thumb in his annihilation of Yemelyanov. Doctors looked it over, but Frazier would not allow them to X-ray it. "I figured if [an X-ray] showed it was bad, I'd be thrown out of the finals," Frazier said. Three inches taller and ten pounds heavier than Frazier, Huber backpedaled from the onset of the bout as Frazier plowed ahead and launched wild left hooks. Only three of the twenty he threw in the second round alone landed. Frazier bore in again in the third and final round, yet Huber remained out of range until deep into the round, when Frazier stunned him with two left hands to the head. Even with the esoteric scoring system embraced by Olympic boxing, it seemed inconceivable that Huber could walk away with the decision. But it was far closer than it should have been. Frazier won a 3–2 split decision, which prompted Red Smith to observe in the *New York Herald Tribune*: "Robbery was averted only because one Argentinian judge, one Finn and one Japanese out-voted a nudnick from Fiji and a Rumanian schmo who thought they were watching a footrace." Buster extended his good hand and said, "Congratulations, big Joe!" An exhilarated yet weary Frazier told the press, "All I intend to do is go home and take it easy for a while."

Five hundred people showed up at Philadelphia International Airport in late October to welcome back Frazier and eight other local Olympians, a rowing team of six oarsmen and two track-and-field athletes. A band played "The American Conquest March." Florence was there with their three children. She told reporters that she was "a little scared," and not sure if she wanted her husband to continue boxing. Frazier was also unsure of himself, if only because no one had yet stepped forward with a deal for him to turn pro. Unlike Olympic champions who would follow him years later, including far lesser talents upon whom big bags of money in endorsements and other commercial arrangements were heaped, Frazier found himself in a state of limbo. X-rays of his thumb performed in Tokyo before he boarded a plane back to Philadelphia revealed a break that doctors stabilized with a temporary cast that enabled him to travel. The injury would later require two operations.

Still reeling from the September swoon of the Phillies, which had come on the heels of the riots and steeped the city into a catatonic state, Philadelphia delighted in the emergence of "The Slaughterhouse Kid." Only twenty, he carried himself in public with an agreeable humility that Stan Hochman characterized in the *Philadelphia Daily News* as "doorman-like politeness." Citations and plaques were handed out to him at banquets, where he stood in a coat and tie for group photographs with other area Olympians. One by one, they signed his plaster cast, which Frazier said he planned to keep in his trophy case. On December 1, Frazier joined 108 members of the U.S. Olympic Team in Washington for a luncheon at the White House, where President Lyndon Johnson extolled the assembled athletes for their haul of 150 medals and observed: "Perhaps I will be forgiven if I note that 'The Star-Spangled Banner' was played so often [at medal ceremonies] that people in Tokyo went around humming it like the number-one hit tune of the day." Four days before Christmas, Frazier spoke for over an hour to the eighth-grade students at Newcomb School in Pemberton, New Jersey. W. P. Moran, their teacher, wrote in a letter to the *Philadelphia Inquirer*: "He was delightful. The children loved him. He was good for them. . . . There should be more Joe Fraziers in the world."

Even as he reveled in the acclaim that was showered upon him,

Frazier remained financially strapped to the point that he had become destitute. His injury had prevented him from returning to his
former job at Cross Brothers. In light of the stature he now had as
an Olympic gold medalist, he hoped that his bosses there would appreciate the public relations value he added and find something for
him. But no position was offered, which he later remarked upon in
interviews as the cause of some dejection. To supplement Florence's
small salary at Sears, a mutual friend encouraged him to approach
former Olympic and heavyweight champion Floyd Patterson for
a small loan. Patterson said he would think it over and call back.
When he never did, Frazier carried a grudge that spilled into the
pages of *Sports Illustrated* a year later. He said, "I had a wife and
three bugs [children] and there was no food in the house. When
that happens a man does a lot of things he don't feel right doing. If
Patterson didn't feel he could afford to loan me a little bread, well,
that was all right. But he could have wrote and said, 'Sorry, but no
go.'" Frazier looked forward to the day they met in the ring.

Too proud to ask for any form of charity that would subtract
from his manhood, he became the beneficiary of a civic expression of it when the local papers discovered he was down to the
lint in his pockets just prior to Christmas 1964. Immediately, help
began pouring in. Mayor James H. J. Tate offered him a temporary job with the Department of Recreation. Seven or eight other
places called with jobs. Frazier jotted them down on a pad. A check
in the amount of five hundred dollars came from John Taxin, the
owner of Old Original Bookbinder's seafood restaurant. Smaller
sums came from others. When the donations were all counted,
they came to twelve hundred dollars. And the toys! "They've been
coming in all day," said Frazier, who arranged them under the tree
until Christmas morning for the children. Overwhelmed with appreciation, Florence would say, "It was the merriest Christmas that
there ever has been."

CHAPTER THREE

CLOVERLAY

Cloverlay. *Philadelphia Bulletin*

When he looked back years later, Joe Hand considered it a lucky break that he ended up in the subway. But it seemed far from that when he was still a young cop in the Philadelphia Police Department and the teletype came in that he had been reassigned to the transit division over an altercation he had had with a prisoner. The subject had been a marine who had just arrived on furlough from Camp Lejeune with a carload of his buddies and, as marines are sometimes apt to do with a weekend off and pay in their pockets, they got drunk and began tearing apart a bar in the Northeast section of the city. The cops were called in to break it up and hauled the besotted bunch away. At the Twenty-Fifth District, the six prisoners became so rowdy in the back of the wagon that it nearly tipped over. One of the arresting officers called out to Hand, "Help me get these guys out of here." Hand had stopped by to give the keys to their only car to his wife, Margaret, herself a policewoman who had just clocked out. As Margaret walked by the rear of the wagon, the marine yelled, "Hey, honey!" and hurled an obscenity at her.

Wiry in build yet hard-nosed, Hand knew he had to do something. "Other cops were standing there looking at us," he said. "Either I handled it or word would get around and the men on the force would start pinching her butt. The department had eight thousand men and only thirty-five policewomen then." Hand walked to the wagon and slid the back door open. The marine eyed him. "Come on out," Hand told him. "Nothing'll happen to you." When the marine stepped down onto the bumper, Hand whacked him in one knee and then the other. Both kneecaps were shattered in a working-over that would also leave the marine swollen with cuts and contusions. When Hand came in the following day to accompany the marine to appear before the magistrate, the desk sergeant gave him a worried look and said, "Is that your prisoner, Joe? My God, they had to take him to Philadelphia General. He may die."

For thirty days Hand reported to the roll call room and sat there as the marine recovered and disciplinary action for excessive force was considered. Finally, Hand was sent down to the subway patrol at Eighth and Market Streets, where the sergeant in charge told him, "We get all the garbage down here." Hearing himself referred to in that fashion did not sit well with Hand, nor did the assignment itself. "They more or less said, 'We'll fix you. We'll send you where we can keep an eye on you,'" Hand said. "But I had the balls to say, 'I want to see the police commissioner.'" Hand had not yet become acquainted with Frank Rizzo, but the two shared a common background. Both were sons of cops—in the case of Hand, a detective who passed away when Hand was twelve. Rizzo looked up from his desk and said, "Come on in, I know what happened. Now, what's your story?" Hand told Rizzo his side of it and claimed that the penalty he had received was too harsh. Rizzo replied, "You nearly killed the guy." He then added, "Look, stay in the subway. Serve your punishment like a man. And come back to see me in a year." Rizzo said he would then assign him to any duty he pleased.

Hand discovered that the subway detail was not as unappealing as he had expected. He worked midnight to 8 A.M. and even again encountered the now-sober marine, who apologized to him for insulting Mrs. Hand. With the exception of a tunnel fire in which Hand rescued two women, there was a low incidence of drama in the subway, where the chief duty entailed hauling drunks off the trains. To stave off boredom, Hand occupied himself by reading the four papers in town. In paging through the sports section of the *Philadelphia Daily News* one day, he came across an article that announced that a consortium of investors had formed to lend support to a young Philadelphia boxer who had won the Olympic gold medal. While Hand was not an avid boxing fan, the undertaking appealed to his entrepreneurial instinct. For a public servant earning thirty-five hundred dollars a year, the buy-in was high—$250 a share. But he asked to buy a share in a letter to the organizer of the corporation, Dr. F. Bruce Baldwin, who Hand remembered had lived on the same street as he did as a boy. Dr. Baldwin wrote back: "If I can't help a fellow Jackson Streeter out, I can't help anybody." Hand withdrew $250 from the credit union and thus became an early member of Cloverlay—an amalgam of the words "clover," which stood for good luck, and "overlay," which was a British term for "a good bet." But Hand would become more than a passive investor. Given the hazards that were then afoot in boxing, Baldwin appreciated the advantages of having a cop on the beat.

—

Whatever else Muhammad Ali would become during his extraordinary journey across the world stage, incarnations that would include champion prizefighter, goodwill ambassador, and pillar of hope for the dispossessed, he began it playfully as a consummate showoff who dreamed only of piles of cash and pretty girls. Growing up in Louisville, Kentucky, the young boy then known as Cassius Marcellus Clay Jr. had such an insatiable appetite for attention that he used to run alongside the school bus instead of taking a seat on it. In a 1964 *Playboy* interview, he told Alex Haley (later the author of *Roots*): "All the kids would be waving and hollering at me and calling me nuts. It made me somebody special." Quickly, it became clear to him as he began his ascent in the ring that "grown people—the fight fans—acted just like those school kids." Coaxed by his amateur coach Joe Martin to overcome his fear of flying and attend the Rome Olympics in 1960, the then– light heavyweight became a whirlwind of sound and color that attracted a parade of enchanted followers in the Olympic Village. With the same alacrity with which he dispensed autographs and mooned over American sprinter Wilma Rudolph, he engaged the press with an originality and charm that belied his average IQ and low class standing at Central High School (376th of 391). When a Russian journalist probed him on the ordeal of blacks in the United States, the eighteen-year-old snapped: "Man, the U.S.A. is the best country in the world, including yours. I ain't fighting off alligators and living in a mud hut." In a strategy to gain attention that was then unique in sports but would become commonplace, he recognized that there was box office in a big personality.

Until Clay appeared on the scene and gave it new energy, boxing was looked upon, with justification, as a province of hoodlums. Captured superbly by screenwriter Rod Serling in the 1962 film *Requiem for a Heavyweight*—in which Clay had a cameo role that paid him five hundred dollars—the grimy underbelly of the sport was pinned down by the character Maish Resnick, the unscrupulous manager of an over-the-hill heavyweight. With a cynical sneer, Maish scoffs: "Sport? If there was headroom, they'd hold these things in sewers." From World War II until the early 1960s, boxing was in the stranglehold of the mob, which strong-armed talent, fixed bouts, and extorted kickbacks. At the head of this shadowy enterprise was Frankie Carbo, a soldier in the Lucchese crime syndicate who was known as "Mr. Gray" because of the gray fedora that sat above his dead eyes. He used fear to leverage a blind interest in the top heavyweights of the era—including Jersey Joe Walcott, Rocky Marciano, and Sonny Liston—and he coerced middleweight Jake LaMotta into taking a dive against Billy Fox by assuring him the title shot he had been denied. In collaboration with James Norris, who owned Madison Square Garden and controlled television and site fees by way of the International Boxing Club, Carbo remained an entrenched force in boxing until 1961, when he, Philadelphia underworld figure Frank "Blinky" Palermo, and three others were prosecuted by U.S. Attorney General Robert F. Kennedy and convicted of trying to chisel in on the earnings of welterweight champion Don Jordan. Upon sentencing Carbo to twenty-five years, U.S. District Judge George Boldt called the so-called "boxing czar" a "menace to humanity and a hardened, degenerate criminal."

Eleven wealthy Kentuckians chipped in a total of twenty thousand dollars to get Clay up and running as a pro in October 1960. With ties to tobacco, whiskey, horses, communications, and banking, the consortium called itself the Louisville Sponsoring Group (as opposed to using the term "syndicate," which had unsavory connotations). Clay would call them his "eleven white millionaire managers." Overseeing the operation was Bill Faversham Jr., an ex-actor who was a vice president at Brown-Forman Distillers Corp. (the producers of Old Forester, Early Times, and Jack Daniel's whiskeys). Beyond the unspoken hope of splitting up a big score, they had a paternal affection for Clay, even as they seemed to agree that he could be "a difficult young man." Chiefly, their aim on his behalf was to shield him from the tax problems that plagued Joe Louis and serve as a firewall between Clay and any encroachment of organized crime. The original contract Clay signed three days before his pro debut on October 29, 1960, called for him to receive a ten thousand dollar signing bonus, a salary of four thousand dollars a year, all expenses paid, and a 50-50 split of all purses. Moreover, the group set up a trust fund in which Clay deposited 10 percent of his earnings and hired veteran trainer Angelo Dundee, who had a spotless reputation and would go on to work with fifteen world champions. Although Clay more or less conditioned himself, Dundee would help him develop his jab, become a savvy presence in his corner, and play a key role as an inexhaustible advocate on his behalf with the press, which in those early days remained unconvinced that the so-called Louisville Lip possessed the cut of a champion.

But they did know the precocious young man could talk. Given an audience of more than one, he would step out of the deep silences that tended to envelope him in private and launch into monologues that were at once wildly comical and highly purposeful. In creating a commercially viable version of himself that he could project upon the world, he borrowed from an unlikely tandem of progenitors. From Sugar Ray Robinson, he acquired a certain elegance of style, which revealed itself in a cool artistry inside the ring and a regal conceit outside of it. But the underpinnings of his seemingly spontaneous showmanship were pure Gorgeous George, the preening pro wrestler with a head of golden ringlets who wore a robe of orchid brocade and blew kisses into the perfumed air as a frenzied crowd heaped scorn upon him. From the very beginning, Clay divined that if it was good to be loved, it was even better to be hated. To that end, he whipped up self-infatuated rhymes—"They all must fall, in the round I call"—and speared his opponents with ugly taunts that included denigrating them as Uncle Toms. In the manner of an unruly child, he equated them with animals or invoked some other inanity, just as he would years later when he hammered away at Frazier by calling him "The Gorilla." Some of it was just playful, some of it had a personal edge, but none of it was done without design. Clay told *Playboy*, "People would start hollering, 'Bash his nose!' or 'Button his fat lip!'" Whatever they called him, Clay professed to be fine with it as long as it put asses in the seats.

None of the juvenile antics Clay unfurled were particularly amusing to Charles "Sonny" Liston. Clay anointed him "The Big Bear"—"a big, ugly one"—and irritated him so thoroughly that Liston once flipped a copy of *Time* magazine on the ground, undid his zipper, and urinated on the illustration of Clay on the cover. Born the twenty-fourth of twenty-five children in Sand Slough, Arkansas, to a cotton farmer who Sonny said "gave me a licking almost every day," Liston joined his mother in St. Louis when he was thirteen, only to end up a street thug who committed serial muggings for small change and participated in an armed robbery of a restaurant, which landed him in Missouri State Penitentiary at Jefferson City just before his eighteenth birthday. He began boxing when a Catholic priest in charge of the prison recreation program laced a pair of gloves on him. Upon his parole two years later, he began entering amateur tournaments, won the National Golden Gloves Championship in 1953, and turned pro under St. Louis Mafia underboss John Vitale. Three years later, he ended up in the St. Louis Workhouse for just under a year after he shattered the leg of a cop during an altercation over a parking citation. Control of his contract passed from Vitale to Carbo and Palermo upon his release, at which point he joined Palermo in Philadelphia. According to Cosa Nostra defector Joseph Valachi, Chicago kingpin Salvatore "Sam" Giancana also had a 12 percent piece of Liston until he was squeezed out of his action in a power play by Carbo and Palermo.

Stamped into his countenance by the troubles he had seen was an expression of raw hostility. Joe Flaherty captured it exquisitely in a 1969 *Esquire* piece when he observed of Liston: "His glacial eyes have all the warmth of an army-camp madam a week before payday." George James worked with Liston as a young trainer in Philadelphia and remembered that he had a violent temper. "When he got that goddamn gin in him, he turned into an animal," said James. Sober, Sonny remained an unnerving presence. Upon hearing the howling objection of a hooker in England that she had only been paid ten quid (twenty-four dollars) for wall-to-wall, floor-to-ceiling sex, he told the promoter who had lined up the assignation and had intervened on her behalf for a pay bump, "What about all the sandwiches she ate?" Whenever he roamed, there seemed to be a patrol car in his rearview mirror. Word circulated in Philadelphia that the police kept his photo pinned to the sun visor of their cruisers. The cops picked him up for loitering on a street corner; the charge was later dropped. They picked him up again for, of all things, impersonating a police officer when he stopped a woman on the Schuylkill Expressway and beamed a flashlight at her as she sat in her car; that charge and sundry others were dropped when the woman accepted his apology. James remembered being pulled over once with Sonny in the passenger seat.

"We saw the police lights behind us and Sonny said, 'George, they just want me,'" James said. "The cop walked up to the car, saw Liston sitting there, and ordered him outside. Hands up on the hood. They searched him up and down, spun him around, and searched him again. Sonny never said a word. They told him to get back in his car and they drove off. Sonny told me that trouble just always followed him."

Sonny could not escape the scorn of society as he ascended to number one in the annual *Ring* magazine rankings in 1960. Even when Liston apparently severed his connection to organized crime and appointed George Katz as his manager the following spring, the sullen ex-jailbird remained unworthy of challenging church-goer Floyd Patterson for the heavyweight championship. President John F. Kennedy joined an anti-Liston crusade by saying that if Patterson skipped over Sonny, "we would all be better off"—including his brother Bobby, the attorney general, then deep into an investigation of organized crime. Even the National Association for the Advancement of Colored People shunned Liston because they considered him to be a poor example for black children. But whatever initial qualms Patterson had that prevented him from giving Liston a title shot were ameliorated when Sonny switched his managerial allegiances to Katz. In what he called the spirit of fair play, Patterson agreed to tangle with Sonny on September 25, 1962, at Comiskey Park in Chicago. Seizing upon the innate decency that this act of liberality reflected in Patterson, Howard Cosell asked Liston what he thought of "this great champion." Sonny fixed his baleful glare upon Cosell and replied, "I'd like to run him over with a truck."

Of the more than six hundred correspondents who poured into Chicago for the fight, including literary figures such as Norman Mailer, Budd Schulberg, and James Baldwin, nearly all found themselves on the horns of a dilemma: they loved the self-probing Patterson but saw no earthly way that he could beat Sonny, who said he would be "ashamed" if Patterson was still standing in the sixth round. With a twenty-five-pound weight and thirteen-inch reach advantage, Liston need not have been unduly concerned. He flattened Floyd with a left hook at 2:06 of the first round. As Patterson slipped out of Chicago in disguise to shield his shame, the sports pages were swathed in black crepe. *New York Post* columnist Murray Kempton tapped out a dirge that echoed the revulsion that had spread across the land when he observed that Liston was "the first morally inferior Negro I can think of to be given an equal opportunity." In light of the annihilation that had occurred, no one could seem to summon even a creative argument for a rematch other than Mailer, who, strafed by unchecked boozing and a lack of sleep, emerged from his bunker at the Chicago Playboy Mansion and proclaimed that Patterson had been beaten because the Cosa Nostra had cast upon him a Sicilian evil eye. But there would indeed be a rematch less than a year later, on July 22, 1963, at the Las Vegas Convention Center, and the outcome was just as cruelly conclusive. Liston floored Patterson three times, polishing him off with a left uppercut at 2:10 of the first round. Next up (if the precocious young blabbermouth from Louisville dared): Clay.

Spasms of visceral fear were triggered by the emergence of Liston. White America looked upon him as the avatar of black rage. To his opponents, he was large and unrelenting and, in the words of *Philadelphia Evening Bulletin* columnist Sandy Grady, had fists as "big as oil cans." Few in the history of boxing had a jab as concussive; Dundee equated it to being clobbered by a telephone pole. In keeping with his stratagem of fomenting chaos, Clay provoked Liston as he was shooting craps at the Thunderbird Hotel prior to the second Patterson bout. Idling in the vicinity as Liston crapped out, Clay chirped: "Look at that big ugly bear; can't even shoot craps." Liston glowered at him, picked up the dice, and crapped out again. Clay chirped once more, "Look at the big, ugly bear. He can't do nothing right." Liston strolled over to Clay, eyed him menacingly, and said, "Listen, you nigger faggot. If you don't get out of here in ten seconds, I'm gonna pull that big tongue out of your mouth and stick it up your ass." Clay was jarred by the encounter—genuinely so—yet he continued to taunt Liston in the ceremonial prefight introductions of ringside celebrities. Dressed in a tailored suit and narrow tie, he ducked through the ropes, nodded at Patterson, and started for the opposite corner, where Liston glared at him from beneath the hood of his robe. With a theatrical flourish, Clay froze, feigned an expression of horror and waved off Liston with a dismissive hand as he exited the ring. Whatever fear Clay had of Liston would be counterbalanced by his calculation that when faced with a growling bear, the way to subdue it would not be by battling it hand-to-claw but by using a psychological arsenal to bring it down to size. And step over it.

Hysteria engulfed the buildup to the February 25, 1964, bout at Miami Beach Convention Hall. To agitate Liston, Clay purchased a bus, affixed the words LISTON MUST GO IN EIGHT on the side of it, and parked it at 3 A.M. outside the Denver house Sonny had purchased once he decamped from Philadelphia, which occasioned the legendary gibe by Liston: "I'd rather be a lamppost in Denver than the mayor of Philadelphia." With the press in tow, Clay shouted, "Come on out here, I'm going to whip you now!" Clay told Liston he was going to drag him home and use him as a bearskin rug. "He drove Sonny crazy," said Jimmy Ellis, who grew up with Clay in Louisville, worked as his sparring partner, and lost to him years later in their only bout. At the weigh-in on the morning of the event, Clay came bounding through the double doors into the room screaming, "Float like a butterfly, sting like a bee! Tell Sonny I'm here!" Clay became so animated with chatter and bombast that his heart rate and blood pressure soared (to 120 and 200/100, respectively). To Dr. Alexander Robbins, the chief physician for the Miami Beach Boxing Commission, Clay appeared to be "scared to death." The doc even talked of canceling the bout, in fear for Clay's safety. Given the carnage that Liston seemed certain of wreaking, that seemed to be a prudent call in the eyes of the assembled press, some of whom were candidly concerned that Clay would end up on a slab in the morgue. *Los Angeles Times* columnist Jim Murray joined in that groundswell of opinion when he whimsically observed that he hoped Clay "clots easily." Darkly, Liston added, "My only worry is how I'll get my fist outta his big mouth."

Only 8,297 spectators were intrigued enough by the apparently uneven pairing to show up at Convention Hall, with six hundred thousand more taking in the event via 371 Theater Network Television hookups. Two inches taller and eight pounds lighter than

Liston, at six feet three inches and 210 pounds, Clay stepped into the ring as a 7–1 underdog, which only cemented the prevailing wisdom that Sonny would devour him with the ease he would expend on the slab of corned beef that had been set out for his victory party at the Fontainebleau Hotel. But what no one expected was that, even as he carried his hands so dangerously low, Clay would be an unhittable target by virtue of his incomparable speed. From the opening bell, Clay was up on his toes and on the move, as Liston lunged forward and swung wildly in search of a quick knockout. The impeccably sculpted Clay was a blur of spontaneity, his legs operating in sublime orchestration with his hands. Clay swarmed over Liston in the third round and opened up a deep cut under his left eye. But just as it appeared that Clay had command of the affair, Liston waylaid him with an alleged act of chicanery.

On his stool between the fourth and fifth rounds, Clay was beset by a sudden burning sensation in his eyes. Liston had been overheard by *Philadelphia Daily News* columnist Jack McKinney ordering his corner to "juice the gloves" with a powerful astringent; Dundee claimed it was Monsel's Solution, which had been used to close a cut under Liston's left eye and was inadvertently transferred to Clay. Rubbing his eyes furiously, Clay held up his gloves and ordered Dundee to "cut 'em off." Dundee replied, "This is the big one, Daddy. Let's not louse it up." Unaware that referee Barney Felix was on the verge of awarding Liston a TKO, Dundee shoved Clay out into the center of the ring. To buy time until his vision cleared, Clay held Liston at bay extending his left hand in his face in a maneuver assistant trainer Bundini Brown called "yardsticking." By the sixth round, he was down off his toes and throwing leather. Liston trudged back to his corner at the end of the round and sat down on his stool, his seconds huddled over him. When Liston did not answer the bell for the seventh round—in reply to the cries of "Fix!" he claimed he had injured his left shoulder— Clay flew across ring in uncontained joy and shouted into the microphones that were shoved in his face: "I shook up the world! I shook up the world! I shook up the world!"

Clay did it again the following day in a way that would be profound. For weeks, it had been rumored that he had become a member of the Nation of Islam—pejoratively known as the Black Muslims. Looked upon by outsiders as a subversive organization in the guise of a religion, the NOI preached a doctrine of separatism from white oppressors. The *Miami Herald* carried a piece that quoted Cassius Clay Sr. as saying that Clay had joined the sect at eighteen. "They have been hammering at him ever since," said the elder Clay, who added that his son Rudy also had been involved with the NOI. Concerned that the revelation would imperil attendance, promoter Bill MacDonald had threatened to cancel the bout if Clay did not publicly disavow his affiliation with the group. Clay balked. On the advice of Malcolm X, the NOI national representative who was then under a ninety-day suspension by leader Elijah Muhammad for saying that the assassination of President Kennedy had been a case of "the chickens coming home to roost," Clay would not comment on his association with the NOI until the press conference on the day after he vanquished Liston. While he said he harbored no hate in his heart for the white man, if only because "I would be nowhere today without the white man's money," he claimed he adopted Islam in search of the peace he had not found "in an integrated world." He said, "Why do I want to live in the white man's way? Why do I want to get bit by dogs, washed down a sewer by fire hoses? Why does everyone attack me for being righteous?" In keeping with his religious conversion, he announced that he would no longer answer to his "slave name"—Cassius Marcellus Clay Jr.—but would henceforth be called Cassius X, later to be renamed by Elijah Muhammad as Muhammad Ali ("most high" and "worthy of praise").

At Ali-Liston II, in out-of-the-way Lewiston, Maine, on May 25, 1965, unfounded rumors swirled that Ali was the target of an assassination plot in retaliation for the slaying of Malcolm X, whose friendship Ali had abandoned after Malcolm defected from the NOI and who had been cut down in February in a hail of gunfire in Washington Heights by underlings of Elijah Muhammad. Originally, the bout had been scheduled for the previous November at Boston Garden. To the chagrin of Liston—who had been established as a 13–5 favorite and had whipped himself into top shape—it was postponed when Ali was forced to undergo surgery for a strangulated hernia. Amid heightened concerns that the promotion had a connection to organized crime, Massachusetts blocked the rescheduling of the event, at which point it was moved north to the tiny Central Youth Center in Lewiston. With the FBI and the bow-tied security arm of the NOI, called the Fruit of Islam, on hand to guard him, Ali dropped a right hand on Liston in the first round that caved him as if he were a ceramic piggy bank. As Liston lay sprawled on the canvas with his arms outstretched, felled by what was called "the Phantom Punch," no one in the sparse crowd of 2,434 or the assembled press was certain what they had seen: Had Ali hit him or had Sonny gone in the tank as part of some betting coup? "Hit he was," confirmed Larry Merchant in the *Philadelphia Daily News*, "but how hard he was hit remains in the realm of occult guesswork." Chaos erupted when referee Jersey Joe Walcott, himself a former heavyweight champion, attempted to steer the exuberant Ali to a neutral corner as the timekeeper pounded out the ten-count on the ring apron. When Liston struggled to his feet, Walcott signaled for the bout to continue, at which point he was summoned ringside by *The Ring* magazine publisher Nat Fleischer and told that the count had reached twelve. With Ali again throwing punches at Liston, Walcott then stepped between the two and declared Ali the winner. Cries of "Fix" were heard across the land.

To the world over which he held dominion, Ali proclaimed himself "The Greatest," and that he would be in the eyes of many, in ways that not even he could then know. Less than two years had passed since the Kennedy assassination, and the "Burn, baby, burn" sixties would soon swing into high gear, as a counterculture sprang up in protest of the Vietnam War and the spark that was ignited in Harlem, Jersey City, and Philadelphia in 1964 became a ten-alarm blaze. Ali soon found himself at the center of it, at once a figure of intense adoration and equally passionate scorn. From the vantage point of his boxing career, he had given the sport the vibrant new face it urgently needed. And yet he himself had something to prove. Given the irregular outcomes in Miami and Lewiston, Ali had been deprived of the unblemished eminence to which he aspired. That would only come years later when he walked through hell with Joe Frazier.

———

With a pail of patching cement in his one good hand, Rubin Frazier scaled a ladder propped up against the side of his house. Granddaughter Lisa Coakley, then five years old, was standing at the bottom holding it when a car pulled up in the yard. Out of it stepped an insurance agent, who had come to collect his monthly premium. "That was how they did it back then," Lisa said years later. "You paid the man when he came around and he recorded it in his big book." She would remember it was the fall of 1965.

"What are you doing on that ladder, Mr. Frazier?" the insurance agent asked.

From halfway up, Rubin replied, "The roof is leaking. I have to fix it." He climbed down and joined his visitor in the yard.

"I need to talk to you," Rubin said. "I need to buy more insurance."

"Why would you need to do that, Mr. Frazier?"

"Well," Rubin said, "I am a very sick man."

Rubin was only fifty-three years old and had been in declining health. Two years before, he had had surgery for lung cancer, and he had battled high blood pressure and diabetes. But he had lived to see his boy win an Olympic gold medal, and he swelled with pride when Billy came back to Beaufort to show it off. Pastor Kenneth Doe was only a small boy back then, but he remembered how the news of what Frazier had accomplished had been met with genuine disbelief. Doe said, "People were like, 'Billy did *what*?' It was as if he had walked on the moon."

When word reached Billy that his father was gravely ill, he gassed up his car and sped down to Beaufort. He arrived too late to say good-bye. At the Second Mount Carmel Baptist Church, an overflow crowd showed up for the funeral, where Rubin Jr.—Jake—performed a gospel solo. On the way to the Habersham Cemetery, they parked for half an hour across from the house in a final farewell. No one who had come to mourn Rubin that day shed more tears than his youngest son.

Grief lingered within Frazier for a prolonged period. But in Yank Durham, he had not just a manager but a surrogate father. Frazier did not have to look far to see Rubin Sr. in Yank, who even cooked up his own corn liquor concoction and sold it on the side. Frazier remembered in his autobiography that he used to deliver packages door-to-door for Yank. While they were all business at the gym, where Frazier was the obedient listener that Durham had longed for, they were partners in pursuit of a good time away from it. Together, they were known to be dedicated womanizers. They squirreled away cash in a secret safe-deposit box they shared, and even had a matching pair of gold-plated pistols made. George James remembered how he used to shoot dice with the two of them at the gym. "They were like brothers," said James. They were so close that whenever Durham referred to Frazier in the press, he would do so in first-person singular. Instead of saying *Joe* needs some time off before *he* fights again, he would say, "I need some time off before I fight again." For his part, Frazier echoed in public whatever Durham had impressed upon him in private and was very seldom in disagreement with him. When he was, Frazier would only walk away without a word.

"I've never come across a fighter who was closer to his trainer, and a trainer who was closer to his fighter, than Joe and Yank were," said Joe Hand. "If Yank told Joe to jump in front of a train, Joe would jump in front of a train."

Frazier found it hard to get going as a pro. In search of backing that would provide him with even a small weekly salary that would enable him to quit the job he then had working on a moving van, he was irked when none of "the dozens of inquiries" he and Durham fielded panned out. "I always said, 'There's gotta be a way, man. I gotta keep goin' somehow,'" Frazier later told *Playboy*. Only Boston sportsman Peter Fuller overlooked Frazier's lack of size and stepped forward with a serious proposition. A car dealer who once managed heavyweight Tom McNeeley and who owned a stable of horses that would one day include 1968 Kentucky Derby winner Dancer's Image (who was subsequently disqualified for having an illegal drug in his urine), Fuller had grown enamored with Frazier in Tokyo and courted him there. But Frazier balked at the arrangement Fuller envisioned: he would have to move to Boston and Durham would have to step aside. With his hand now healed, Frazier began looking toward his pro debut, prior to which he engaged in some verbal sparring with Liston and Ali on a telephone hookup at a Philadelphia restaurant to promote the closed-circuit telecast of their rematch.

Frazier told Liston he heard he had been going easy on his sparring partners.

Liston said he "sent one fellow home" just the other day. He invited Frazier to come up for the fun.

Frazier asked Ali, "Do you have any advice for me?"

Ali replied dismissively, "Yes. Lose some weight and turn light heavyweight." He then told Joe, if the commissioner would allow him to take on three challengers in one night, "you might be one of them."

Leotis Martin worked with Frazier at the Twenty-Third PAL in the spring of 1965. The young light heavyweight was employed as a sheet steel pressman, had occasionally sparred with Liston, and also fought under Durham, who Hand remembered had an innovative way of keeping Martin focused. "Yank would use a clipper and sharpen his fingernail to a sharp point," said Hand. "When Martin sat down in his corner between rounds, Yank would stick him as he was getting his mouthpiece out and tell him, 'Don't come the fuck back here again.' In other words, knock him out and you won't have to go through this shit again." Frazier sparred with Martin as the latter prepared for his bout at the Arena in Philadelphia with Lucien "Sonny" Banks, a journeyman from Detroit who, in a losing effort in 1962, had earned some small fame when he became the first pro opponent to floor Clay. Banks outweighed Martin by twenty-five pounds and had him in trouble in the ninth round when Martin dropped him with a right hand to the temple that traveled no more than six inches. As Banks remained in a semiconscious state on a table in his dressing room, Martin sang the praises of his sparring partner, of whom he said: "Joe Frazier is as hard a hitter as anybody I ever fought." Banks died two days later of brain injuries at Presbyterian Hospital. He was twenty-four.

For a young boxer with promise, opponents are lined up with an eye toward expediency: get in, get out, and move on. In the case of Frazier, he was paired up for his pro debut at Convention Hall in Philadelphia with Don Hobson, who had lost six of his seven bouts and had fought just once in two years. But Hobson begged off with a sprained ankle. To take his place, promoter Lou Lucchese found Roy Johnson for what would also be his pro debut. But Johnson did not show up. Scrambling, Lucchese buttonholed Elwood "Woody" Goss in the lobby of the Benjamin Franklin Hotel. Goss shrugged. "I said, 'Well, he's got two arms like anybody else.' I thought if I got lucky I might tag him." Goss did not get lucky. Frazier warmed up longer in his dressing room—two rounds—than he spent in the ring. He stunned Goss with a left hook and battered him for an additional twenty seconds until referee Zack Clayton stopped the fight at 1:42 of the first round. Clayton said he did so to keep Goss from getting killed. Frazier asked aloud back in his dressing room, "How did I look?" He earned $125.

Beyond a lapse that occurred in his second fight, when Mike Bruce floored him in the first round for an eight-count, Frazier sailed to easy victories in his three remaining bouts in 1965. But boos cascaded upon him at the conclusion of each, if only because his opponents were what Durham called "runners," which is to say they were less than agreeable to mixing it up. Two of them could scarcely tie their shoes and the other did not even have a pair. Abe Davis—who billed himself as "The Hartford Hatchet"—had called Durham the week before his scheduled bout with Frazier at the Hotel Philadelphia Auditorium and warned him that he was "no pushover. You better have Joe Frazier in shape for a fight." But when Davis showed up in Philadelphia, he discovered that he was without his boxing shoes. "Here, you can take mine," Frazier told him. "I can get another pair." When Davis explained that he wore a size 13 triple E, someone was sent in search of high-top sneakers. The pair that was found had tattered laces and were too tight to accommodate sweat socks, so Davis slipped into some olive-green dress socks. Frazier toppled him with a barrage of blows at 2:38 of the first round. In the *Philadelphia Daily News* the following day, there was a picture of Davis stretched out on the canvas with the soles of his shoes exposed. One of them had a yawning hole in it.

———

Invited to appear at the dedication of the Bright Hope Baptist Church, at Twelfth and Columbia in North Philadelphia, Dr. Martin Luther King Jr. ascended to the pulpit on May 2, 1965, before a congregation of fifteen hundred people and an overflow closed-circuit television crowd of two thousand. King had been asked to come that Sunday by Dr. William H. Gray Jr.—also known as Billy Gray—the pastor at the church and a close friend. In keeping with his message of peace, equality, and universal harmony, King pronounced that in the struggle for civil rights "the church should be the one place where men remove their burden of class." At the apex of his soaring plea for unity, he articulated that he could foresee a day when "whites, Negroes, Jews, Gentiles and all men, can join hands and sing the words of that old Negro spiritual, 'Free at last! Free at last! Thank God Almighty we are free at last.'" So stirring were these words when he had uttered them in his 1963 "March on Washington" speech that they would live on beyond his assassination three years later on the balcony of a Memphis motel.

Obsessed in his youth by dreams of playing professional baseball, Gray had given Frazier work as a janitor at the church. "The Reverend Gray was the backbone of the community in North Philadelphia," said Mazie. "He saw to it that women and children had clinics to go to and that the men had jobs if they needed one." It is unknown if Frazier was in attendance on the Sunday that King spoke, but he and Durham were both members of the church. Frazier had an admiration for King that would only later be shared by Ali, who Dr. King had said had become a "champion of racial segregation" when he joined the Nation of Islam. Even if Frazier would concede that there were periods when he lapsed as a churchgoer, he remained true to his Christian upbringing and considered himself a man of God. For inspiration across the years, he would play a cassette tape in his car of sermons by C. L. Franklin, the pastor of the New Bethel Baptist Church in Detroit and the father of the renowned soul singer Aretha. In keeping with the principles espoused by King, Frazier did not think of the white man as the same "incorrigible white devil" as Ali did in those early days. While he was not blind to color, he tended to take people as they came in a down-to-earth way, and not look upon them uniformly in one way or another; if you were good to him, he was good to you. When he came into some money as his career began to flourish, he would write Reverend Gray occasional big checks to help keep the church doors open.

Beyond just giving Frazier a leg up by throwing him some work, Gray played a critical role in helping him secure backing. By virtue of his positions on the Civil Service Commission and the Philadelphia Housing Authority, he used his influence with the top business leaders in Philadelphia to help organize Cloverlay. Intended to aid Frazier in the same fashion that the Louisville Sponsoring Group had supported Clay—which is to say, provide him with seed money and enable him to avoid problems with the taxman—the investment group was headed by Dr. F. Bruce Baldwin, the president of Abbotts Dairies and author of a doctoral thesis at Penn State titled "The Chemistry of Frozen Milk and Cream." Baldwin referred to himself as "just a milkman," but he was more than that, helping out with the Board of Public Education, the Heart Association, and other civic endeavors. Boxing was not an area in which he had any expertise, but his cousin Bowers Baldwin had once been a Golden Gloves champion and he considered himself a buff. Forty investors ponied up $250 for each of the initial eighty shares and included, in addition to Baldwin and Gray, Bruce Wright, an estate lawyer who oversaw contract negotiations; Thacher Longstreth, executive vice president of the Greater Philadelphia Chamber of Commerce; Arthur Kaufmann, the former chief executive of the Gimbels department stores in Philadelphia; Robert G. Wilder, an advertising executive; Harold Wessel, head of an accounting firm, and—if only for the amusing story angle it gave him—sportswriter Larry Merchant. In a droll column in the *Philadelphia Daily News*, Merchant observed that "there is a ring of amateur, wholesome-as-milk goodness to the syndicate" and looked forward to finding a "suitable opponent" for Frazier in his Cloverlay debut.

Gleefully, Merchant added: "I consider anyone suitable who weighs under 160 pounds, has 40-80 vision, size 18 feet, short arms and hasn't been in a gym in two years. We'll moider the bum."

Ham and cheese sandwiches were served at the luncheon that was held at the Bellevue Stratford Hotel. ("I admire our frugality," Merchant observed of his fellow investors. "I'll be damned if I'm going to stuff the fat cats of the press with onion soup and steak.") Under the terms of his initial three-year agreement, Frazier was guaranteed a weekly salary of a hundred dollars ($773.82 in 2017 dollars), all training expenses, and 50 percent of his purses. Cloverlay paid Durham 15 percent of its 50 percent share and accorded him final say over whom Joe fought, when, and for how much. Training expenses ate up another 15 percent, which left the investment group with 20 percent of the proceeds to split among themselves. Cloverlay held two options to extend, with Frazier receiving 55 percent of his purses when the first option was exercised and 60 percent after the third. Frazier called the arrangement "just swell."

Owning a piece of a heavyweight then held a certain appeal to men of means, if only because it provided them with a chance to step out of the gray world of commerce and into the company of sportsmen. Cloverlay afforded them that opportunity. "It was a cocktail stock," said Joe Hand, who came aboard after the initial offering. "They could go to the fights, have a drink or two, and enjoy themselves. It was a night out." Not sold on any open exchanges, shares were available only to Pennsylvania residents, which came as something of a disappointment to George W. Romney, then governor of Michigan and father of future Republican presidential candidate Mitt. "He loved Joe," said Hand. "Over the years, he would call me and say, 'How is he doing? Are you sure you can't sell me some stock?'" Of the forty investors who did get in on the action, a dozen or more hailed Frazier with cheers as he stepped into the ring on January 14, 1966, for his Cloverlay debut. The "suitable opponent" found for him was yet another eleventh-hour substitute, Mel Turnbow, who stepped in when two previously scheduled opponents bailed. Heavier than Frazier by thirty-two pounds, Turnbow had won seven of his nine fights and floored Floyd Patterson in a sparring session. Whatever apprehensions Durham and the syndicate had were dispelled when Frazier clubbed the six-foot-two, 231-pound Turnbow with a left hook in the first round that Frazier said was "so solid my arm ached right up to the shoulder." Turnbow was unconscious before his head bounced on the canvas.

Cloverlay provided Joe with comprehensive care. It saw to his needs inside the ring by providing him with patient management that allowed him to develop his skills, and catered to them outside of it in a variety of ways that were aimed to protect him and enhance his profile. Financially, in addition to providing him with an increasing weekly draw, they placed his money in what were understood to be sound investments and kept his taxes current. To shelter him from the wear and tear of the punishing Philadelphia gyms, they purchased an old building on North Broad Street for ninety thousand dollars and converted it into the Cloverlay Gym, which immediately became the finest facility in the city. Cloverlay engaged the clever New York publicist Joey Goldstein to spread the word on Joe, whose communication skills were then a work in progress. Though Frazier was looked upon by the writers who followed him with a genuine fondness, *New York Times* columnist Robert Lipsyte told me he was a "hard interview," explaining: "It was hard to understand him. I think he tried. He was pleasant." To help Joe with that and perhaps enable him to reel in more endorsement work, Hand enrolled him in an elocution class at Temple University.

Frazier narrowed his eyes and asked, "Elocution class, what's that?"

"Well," Hand said, "they teach you how to speak."

"Fuck you," Frazier said. "I know how to speak."

Hand laughed years later and said, "Jesus Christ, he was so furious, I thought he was going to throw me out the window."

Hand had only attended an occasional bout before he became an investor in Cloverlay. When he joined the group, he remembered that his mother asked, "Joseph, I understand you bought into a fighter." Hand said yes. She then asked, "And he is black?" Growing up in Lawndale, in Northeast Philadelphia, a community that was surrounded by cornfields during the 1930s and '40s, Hand said, "I doubt I even saw a black when I was a kid." Once he was in the fold at Cloverlay, it was not long before Hand was asked to take on an expanded role, which included lining up off-duty policemen to keep an eye on the gate receipts and discourage unauthorized fingers from dipping into the till. As time passed, he took on additional jobs within the organization that included ushering Frazier to and from appointments, during which they spent many hours together and became friends. Hand remembered that he once had Frazier to his house for dinner, where his young daughter stood up and began walking around the table. Hand told her, "Margaret, sit down, would you?" Frazier said, "She is looking for my tail." While he said it with an air of conviviality, it was clear to Hand how acutely aware Frazier was of the chasm that existed between the races. And yet it became equally clear how unacceptable the status quo was to him when the Union League extended an invitation to him to speak.

Frazier said he would be happy to do it.

"Are you sure?" Hand asked.

"Yes," Frazier replied. He said it emphatically.

"Well, you know, they would never allow to you to become a member there," Hand said. "Does that make a difference to you?

Frazier replied, "My hope is when they meet me, they *will* let black people join."

Hand would say years later, "That was his attitude. He knew there was prejudice, but he was trying to overcome it."

Hand's daughter Margaret added, "He thought he could open their eyes and let them see that he was just a man, just like anyone else was. He spoke there. And you know what? They did not change their rule for a long time. But that was always his thinking. Dad took him places where he was not allowed to walk in the front door."

By this point Hand had escaped the subway. Rizzo happened to run into him one day down there, one and a half years into his exile, spotted his badge, and said, "Hand! I thought I told you to come see me after a year." Hand told him he was happy underground. But Rizzo promoted him to detective. Hand became his eyes and ears. "Rizzo liked to know what was going on in the city," said Hand. "Dope was not yet as big as it would become, but you had numbers, horses, and prostitution." Hand roamed the nightscape of Philadelphia, dining in fine restaurants and dropping by seedy bars. No one on the force had better contacts or was a sharper dresser. Hand said, "Rizzo used to come to me and say, 'I have a meeting next week with so-and-so. See what you can find out about him.' And I would come back a week later and know more about who he was meeting with than the person knew about himself." And then one day in 1970 Rizzo called Hand, told him he understood that he was involved with Frazier, and asked, "Do you think Joe would like to sit down with me and have lunch?"

CHAPTER FOUR

ASSWHUPPINGS

Joe and Muhammad Ali, 1967. AP Images

he old Madison Square Garden stood on Eighth Avenue between Forty-Ninth and Fiftieth Streets, in an area that was once known as Jacobs Beach. Along with steering the career of Joe Louis, promoter Mike Jacobs held boxing at the Garden in an iron vise under the aegis of the Twentieth Century Sporting Club, which had its offices a crosstown block away at the Brill Building. Under a big marquee that announced the attraction of the evening in block letters, which in any given week would have included an NBA or NHL game, the circus, or even a political rally, the entrance to the Garden was flanked by a hot dog counter on one side and a hat store on the other. Inside, assorted bookmakers camped out in the shadowy corners of the lobby, where there stood a bronze statue of turn-of-the-century lightweight Joe Gans, "The Old Master." Clouds of cigar smoke befouled the interior of the arena itself, which had a balcony that became a launchpad for projectiles in the event of a controversial decision. Then with the *New York Journal-American*, Dave Anderson remembered that a whiskey bottle flew by his ear when Gabriel "Flash" Elorde, of the Philippines, won an unpopular ten-round split decision over Frankie Narvaez, of New York by way of Puerto Rico. Fire axes, beer cartons, and splintered pieces of wooden chairs descended upon press row, which prompted Anderson and *Journal-American* columnist Jimmy Cannon to crawl under the ring for safety.

Even as Eighth Avenue became overrun with porn shops and dodgy bars during the 1960s, the Garden occupied the center of the boxing universe. Commissioned by promoter Tex Rickard in 1925 to replace the 1890 incarnation of the building—which had been actually located at Madison Square—it was a beehive of matchmakers, managers, publicists, reporters, and assorted sponges. Bob Goodman, the son of Garden publicity man Murray Goodman and later a well-traveled publicist himself, had vivid memories of the pitch-and-yaw of the Garden boxing department on the second floor, which was adjacent to *Ring* magazine headquarters. "Spittoons were everywhere you looked," Goodman said. "No one was without a cigar." Amid the overlapping conversations, United Press International reporter Jack Cuddy sat in a corner chair and pretended to doze, his ear peeled for an angle he could peg a story on. Goodman said, "Suddenly, he would open an eye, look around, and shuffle out the door. The next thing you knew, UPI had a scoop." New York City still had seven dailies until the early 1960s, and each of them had two or three men who pounded the boxing beat. Big play was given to boxing and horse racing in the sports pages in what remained of that golden *Guys and Dolls* era, when men hurried by in snap-brim hats and overcoats with wide lapels, and the women who accompanied them were dolled up in high heels and furs.

New York would embrace Joe Frazier in a way that Philadelphia had not yet done. Notwithstanding the Yuletide generosity that had been showered upon him by the fans there two years before, he would become increasingly irritated by what he perceived to be a lack of appreciation by his adopted hometown. Of his relationship with Philadelphia, Frazier brooded: "They don't know when they've got something." But Madison Square Garden did. With the construction of yet another incarnation of the Garden under way, scheduled to open in early 1968, the search had commenced to identify new stars who could be crowd pleasers in an era in which television had more or less abandoned the sport. Thus, the Garden introduced Frazier on March 4, 1966, on a card of young heavyweights headlined by Jerry Quarry, a hard-nosed counterpuncher from Bellflower, California, who had two essential components that promised certain box office: he was as white as a hospital bedsheet and possessed an inclination to brawl that would have elated his Irish forebears. On that evening at the Garden, Frazier stopped Buffalo cop Dick Wipperman on a fifth-round technical knockout and arguably stole the show from Quarry, who battled the more experienced Tony Alongi to a ten-round draw. For Harry Markson, the president of Garden boxing operations, and Teddy Brenner, his matchmaker, the potential pairing of Frazier and Quarry figured to bring a windfall.

The twenty-two-year-old Frazier stayed busy until the Garden could find a way to book him again. At the Hotel Philadelphia in early April 1966, he ran his record to 7-0 by scoring a second-round knockout over Charley Polite, with a left hook that broke his jaw and sheared two molars at the gum line; Dr. Baldwin was so distressed by the spectacle that he had Cloverlay cut Polite a $250 check for groceries until he could work again. From there, three busloads of fans followed Frazier from Philadelphia to the

Civic Arena in Pittsburgh, where he chopped down six-foot-five, 237-pound Don "Toro" Smith in the second round—again with that left hook. Joe shrugged: "I work harder than that in the gym." Next, he set aside his fear of flying and headed to Los Angeles, where he acquainted two opponents, Chuck Leslie and Memphis Al Jones, with his signature punch and flattened them a week apart at the Olympic Auditorium—Leslie in three, Jones in one. At Convention Hall in Philadelphia in July, he floored Brooklyn barber Billy Daniels with left hooks in the second, third, fourth, and sixth rounds before the fight was stopped in the sixth. With an ice pack held against a bruised eye, Frazier said with a grin as he talked with reporters, "What do you think of that? I have a black eye. I look like a fighter." Although he now had eleven victories in the bank—"asswhuppings" all—he would remain untested until Brenner signed him to face Oscar Bonavena.

They called him "Ringo." From Buenos Aires, he wore Prince Valiant bangs, his hair cascading in curls. With his lantern jaw, outsized neck, eighteen-inch biceps, and two flat feet, Bonavena had the appearance of a caveman, which the Garden used to its promotional advantage by dressing him up for a photo shoot in animal skins. Schooled by seventy-six-year-old Charley Goldman, the legendary trainer who had shaped Rocky Marciano into the undefeated heavyweight champion—and who said of Oscar, "He hits harder than Marciano did"—Bonavena proved to be a disengaged pupil, given his preoccupation with the opposite sex. His handlers had to keep him more or less under house arrest while he was in training. Stan Hochman observed in the *Philadelphia Daily News*: "There are only a few things Bonavena enjoys more than talking to girls and they also involve girls." Roadwork bugged him, as did training in general, yet he endured enough of it to win twenty-one of his twenty-three fights, which had included a ten-round decision over Canadian George Chuvalo as a 9–5 underdog at the Garden in his previous bout. In keeping with his uncouth manner, Bonavena disparaged Frazier by sniffing the air and asking, "What is that smell? Is that you, Joe?"

Close to six inches of rain fell on Manhattan the day and evening of September 21, 1966, the largest downpour the city had seen in sixty-seven years. Still, a crowd of 9,069 pushed through the turnstiles for what Durham had expected to be an easy win for Joe. The ungainly Bonavena impressed neither of them. While the Argentine hit hard, he was a disorganized puncher whose footwork appeared to be encumbered by leg irons. Sandy Grady portrayed the lumbering Oscar with amusing precision in the *Philadelphia Evening Bulletin:* "Mostly, he fights like a guy trying to push a car out of a snow-bank." But Bonavena could be a handful, which Frazier discovered to his chagrin in the second round. With a grunt, he caught Frazier with a left-right combination that sent him folding to the canvas. Joe got up at the count of five. Only seconds later, Bonavena floored him again with a left hook. Frazier bounced up at the count of two, grinning, as if to assure Bonavena he had not been hurt. But he was in jeopardy; Bonavena would have been declared the winner under the "three-knockdown rule" if he had dropped Frazier again. From their seats, 146 Cloverlay shareholders squealed in alarm: "Keep your hands up!" "Slip and move!" Frazier survived the second round, collected himself, and battled Bonavena toe-to-toe across the remaining eight rounds in an affair that deteriorated into a back-alley brawl highlighted by blows beneath the belt. Frazier walloped Bonavena in the groin in the seventh round—he later called it an "accident." "Ringo" answered it with a less-than-accidental blow down under, followed by a wink and a faux apology. At the end of ten exhausting rounds, Frazier was awarded a split decision that drew howls of disapproval from the Garden crowd.

Whatever the Bonavena bout lacked in style points, it was revelatory insofar as Frazier was concerned. Going into it, he was an unknown quantity in the eyes of the press, which could draw no firm conclusions from the parade of cops and barbers he had beaten. But his narrow victory over Bonavena provided a glimpse into the sturdiness of his hide. Frazier proved he could take a punch, get up off the floor, and go ten rounds against a ranked opponent—and win. That said, it was abundantly clear that he still had work to do before he was up to the challenge of Ali, who loomed on the horizon even then as a big payday for Frazier. Within the inner circle of Cloverlay, there were some who would have liked to see it happen sooner than later. Even Frazier seemed eager to get it on with "Clay," yet he only addressed the subject when asked and always with the caveat that he deferred to Durham when it came to picking opponents. When Ali invited Frazier for a showdown on April 25, 1967, in Tokyo, for a rumored purse of $250,000—three days before he would be summoned to appear at the Armed Forces Examining and Entrance Station in Houston—Durham summarily rejected it, later explaining: "If I had sent that boy in there against Clay, I would have been the laughingstock of boxing." Durham calculated that Frazier would still need one and a half years of "college" before he took on "Professor Clay."

History engulfed Ali with the escalation of the Vietnam War. Though he had flunked the U.S. Armed Forces qualifying test in 1964 due to poor writing and spelling skills, his draft status was reclassified as 1-A in February 1966 as the demand for soldiers increased. Famously, Ali told the *Chicago Daily News*, "I don't have any personal quarrel with those Viet Congs." Called to step forward on April 28, 1967, Ali refused induction on the grounds that Muslim orthodoxy forbade him to "take part in no wars unless declared by Allah or the Messenger." Immediately, the World Boxing Association stripped him of his championship and the New York State Athletic Commission suspended his boxing license. A jury of six men and six women convicted him of draft evasion on June 20, whereupon he was sentenced to five years in prison and a ten-thousand-dollar fine. He remained free on bond as his attorneys hurried to file an appeal. In the *Philadelphia Daily News* that week there appeared an article under a headline that asked, CLAY: FRAZIER BOUT BEFORE PRISON?

———

Never one to seek attention for himself, Eddie Futch dwelled in the shadows of his profession for years, his brilliance as a trainer known only to his more discerning peers and true aficionados of the sport. He was the son of a Mississippi sharecropper who had come to Detroit in search of factory work and settled in the Black Bottom neighborhood. Only five foot seven but quick with his hands and on his feet, young Eddie played basketball for Northeastern High School and later suited up for the semipro Moreland YMCA. He began boxing at the Brewster Recreation Center in the same stable with then-amateur light heavyweight Joe Louis. From 1932 until 1936, while Futch worked as a waiter at the Hotel Wolverine, he campaigned as an amateur lightweight for the Detroit Athletic Association. He won the Detroit Golden Gloves in 1933. Although Louis had a forty-pound weight advantage, the Brown Bomber used him as a sparring partner. He told Futch, who had a 37-3 amateur record: "When I'm sharp enough to hit you, I'm sharp enough to hit anyone." When a heart murmur prevented Futch from turning pro in 1936, he began working with a group of young boxers at Brewster that included Berry Gordy, who would go on to found Motown and who once told *Sports Illustrated* that there is "a little bit of Eddie Futch" in Diana Ross, Stevie Wonder, and Michael Jackson.

To support his wife and four children, Futch had a job during World War II as a spot welder at Ford Motor Company. In his early portfolio of pro talent, he had two contenders—middleweight Jimmy Edgar, who lost twice to Jake LaMotta and fought him again to a draw; and welterweight Lester Felton, who defeated Kid Gavilán and Carmen Basilio. By car, he ferried them and others to engagements across the Midwest, up and down the Eastern Sea-board, and into the Deep South, ever vigilant not to allow his fuel tank to drop below half for fear that he would find himself on a country road at a late hour and encounter a gas-station attendant who would deny him service. "I was never one to look for trouble, but I would not take abuse," Futch told me of his journeys through the segregated South. "I had an automatic pistol up under the dash just in case. Thank God I never had to use it." Neither Edgar nor Felton ever fought for a championship. Edgar had problems with his eyes—cataracts—and Felton had problems with a woman. Futch told me, "I would tell him one thing, and she would tell him the opposite." He visited California in 1951 and stayed, at the in-sistence of a fighter he had once trained. Six years later he had his first champion, welterweight Don Jordan, who found himself under pressure by the mob to go in the tank. To ensure Jordan remained out of their clutches, Futch switched hotel rooms with him prior to a bout in New York in order to cross up any "visitor" who happened to drop by in an effort to strong-arm him.

Sure enough, at 3 A.M. there was a knock on the door of the room that Jordan had vacated.

Futch answered it.

Standing there was Sonny Liston.

"What do you want?" asked Futch.

Liston glowered, pondered the question, and replied, "Towels."

Eddie once told me, "The joy I derive from boxing is a simple joy. And that is to see talent blossom and flourish." With a receding hairline atop his furrowed brow, Futch had broad interests that ex-tended beyond boxing to nineteenth-century British poetry, which he discovered as a young teenager and used to assuage the grief caused by the abandonment of his father. Orderly in his habits, he exuded a courteous air and spoke precisely, striving not to end any spoken sentence with a preposition. In a calm and reassuring voice, he worked the corner with a cool professionalism and attention to detail that would bring him eighteen champions. Beyond his wealth of knowledge and experience, he understood that he held human lives in his hands. He had witnessed the death of seven fighters in the ring. One was obscure Detroit lightweight Talmadge Bussey, who, Futch remembered, was handled by two brothers the evening he faced Luther Rawlings in October 1949. When a battered Bussey came back to his corner at the end of the eighth round, the two brothers squabbled over whether to allow him to continue. "One of them wanted to keep him on his stool, while the other tried to send him out for the ninth round," Futch would remember. "The bell rang and they pushed him out. Bussey was hit and he never woke up."

Futch hooked up with Frazier at the invitation of Durham. He had come highly recommended to Durham as someone who could help him maneuver Joe to the championship, given his uncanny aptitude for sizing up styles and choosing opponents. "Yank asked around for people who could help him," said Futch, who was unsure if he could free himself up. He was working as a clerk for the U.S. Postal Service, which he knew frowned on moonlighting, so he scarcely had time to look after his own fighters. But he agreed to take a look at Joe if Joe would come to California. Futch picked up Joe, Durham, and the sparring partner who had traveled with them at the airport and worked Joe out at the Main Street Gym in Los Angeles. Futch liked what he saw, and he became fond of Joe. The two clicked. Over the apprehensions of Durham, Futch lined up the Chuck Leslie and Memphis Al Jones bouts in May and promised an easy outcome in both.

While Futch did not attend the Bonavena bout in New York that September, it was clear to him even on television back in L.A. that the two second-round knockdowns exposed a weakness in Frazier that called for urgent attention. "Against a taller opponent, Joe would walk straight in and get hit," said Futch. "He had to learn to punch out of a bob-and-weave." When Frazier returned to

Los Angeles to prepare for his bout against the experienced Eddie Machen in November, Futch drilled Joe in the fundamentals of the bob-and-weave by stretching a rope between diagonal ring posts and challenging him to duck under it as he advanced on an imaginary opponent. To get some sparring in, Futch hooked up Frazier with Quarry, who also trained at the Main Street Gym. The action was so intense that it had to be stopped when Quarry suffered a cut lip that required nine stitches.

Promoter Aileen Eaton was appalled. "What a foolish thing for the managers to allow," she told Jack Fried, of the *Philadelphia Evening Bulletin*. "It could have been Frazier who was hurt, and our show would have been broken up."

Ringside at the Olympic Auditorium was aglitter with Hollywood stars, including Lee Marvin, Robert Mitchum, and Milton Berle. While Machen had been a former top contender who had been ducked by Floyd Patterson during his championship reign, he was on the downside of his career at thirty-four, his legs no longer able to steer him out of trouble. Still, he had handed Quarry his first pro loss the previous July and, according to Futch, was "someone who could box and punch a little" and give Joe some necessary schooling. Some in the press were unsure if Frazier was seasoned enough to handle the veteran Machen. But he came out in the first round and sent Machen sailing through the ropes onto the ring apron with a concussive left hook to the jaw. Dazed, Machen crawled back through the ropes with the help of the referee, who inexplicably held up the bottom strand instead of continuing the count. Frazier would say later, "Why'd the ref help him? Why'd he help Ed back in?" Frazier poured it on in the ensuing rounds, building a lead in the scoring as he drove Machen into the ropes. But Machen was nothing if not game. In the eighth round, he was backed into a corner when he clipped Frazier with a left hand that would have floored him had he not clutched Machen by the leg. Frazier came back in the ninth round and pounded Machen with such blind fury that Berle stood up at his seat and shouted, "Stop it! Stop it!" The referee obliged him twenty-two seconds into the tenth round. Machen called Frazier "a tough kid [who] should go a long way." Actor Lee Majors, the star of *The Big Valley* and later *The Six Million Dollar Man,* looked into buying Frazier from Cloverlay.

"Lee asked me if I would approach Yank," said matchmaker Don Chargin. "He and Burt Reynolds owned a couple of fighters. Later, they would have the welterweight Andy 'The Hawk' Price, who beat Carlos Palomino, and José 'Pipino' Cuevas. No figures were mentioned; he just wanted to see if Joe was available. So I asked Yank, who laughed and said, 'You want me to sell a gold mine!'"

Under pressure by Cloverlay to agree to a Frazier-Quarry bout, Durham flew Futch to Philadelphia to persuade the board to back away from the proposal, given that it would be a "hard fight" and that neither Joe nor Jerry was well enough known at that point to be worthwhile financially. "This fight is coming, there is no avoiding it—but not yet," Futch told the board. They concurred. Instead of Quarry, Frazier was matched next with Doug Jones, who had come within an eyelash of upsetting Clay at Madison Square Garden in 1963 in an unpopular decision. But that had been four long years before and Jones was now frayed at the edges, still a "name" but in no way more than a stepping-stone for Frazier when they squared off at the Arena in Philadelphia on February 21, 1967. Quoted in the papers as calling Joe "overrated," Jones found himself under a barrage of leather from the opening bell. Heavier by seventeen pounds, Frazier had Jones in trouble with punishing body shots in the third and fourth rounds before Durham instructed him in the corner prior to the sixth to take Jones out with "one shot." Again, the decisive blow was a left hook. While Frazier would claim with the eye of a perfectionist that he had thrown better ones, he launched it out of his bob-and-weave and spun Jones into the ropes. Sandy Grady observed in the *Philadelphia Evening Bulletin* that Jones "froze in midair, one elbow leaning on the ropes like a meditative Martini drinker at a bar." Jones slid to the canvas, where he remained sprawled on his back for five minutes. The outcome appeared to spell the end for Jones, whose manager, Alex Koskowitz, said: "From now on, he fights with his wife." Frazier had taken yet another step closer to "Clay," even if Durham was not hearing any of it.

Ali was then in preparation for his March 22, 1967 title defense against number-one contender Zora Folley at the Garden—his last fight before he would be stripped of his championship. The Champ held a public workout in the basement of the arena. There, according to Dave Anderson in the *New York Times*, he stepped into the ring before a crowd of spectators that included businessmen, dockworkers, college students, and even a priest. With Ali was Sugar Ray Robinson, whom the announcer exalted as "the greatest fighter who ever lived." Ali grinned and corrected: "Next to me." As Ali was having his gloves laced on for his sparring session with Jimmy Ellis, he spotted Frazier standing in the doorway in a plaid jacket.

"Joe Fraaaaaaaaaazier!" he boomed. "Come up here and have a talk with me."

Frazier climbed up into the ring. He looked Ali up and down, smiled, and said, "I thought you were much larger. You look pretty small to me."

Ali replied, "If you even dreamed about fighting me, you'd be in big trouble."

Ali peeled back Frazier's unbuttoned plaid jacket to reveal a pair of suspenders and said, "Those won't keep you standing." Joe laughed. Ali then told Joe, "Two more years."

"I'll be ready," Frazier promised.

With Ali out of action and possibly headed behind bars for who knew how long, the World Boxing Association announced that it would hold an elimination tournament to choose a new champion in conjunction with Sports Action, Inc. Spearheaded by Mike Malitz, Sports Action included elements of Main Bout—attorney

Bob Arum, former Cleveland Browns running back Jim Brown, and Fred Hofheinz, the son of Judge Roy Hofheinz, the mastermind behind the Houston Astrodome—which had been involved with Ali in the theatrical television promotions of his bouts. Sports Action, Inc. was backed by the American Broadcasting Corporation, which enabled Malitz to outspend rival Madison Square Garden and ultimately line up eight of the top contenders in the heavyweight division: Oscar Bonavena, Jimmy Ellis, Leotis Martin, Karl Mildenberger, Floyd Patterson, Jerry Quarry, Thad Spencer, and Ernie Terrell. While it was preposterous to think that the public would embrace any of these aspirants as a legitimate champion in place of Ali—who had by then beaten Mildenberger, Patterson, and Terrell and would later beat Bonavena, Ellis, and Quarry—the upside appeared to be that these would be lively bouts and would perhaps pump some unpredictability back into a sport that been dominated by Ali. ABC slotted the bouts as programming for *Wide World of Sports* beginning August 4, 1967, in Houston.

Cohorts of Leotis Martin strolled the sidewalk with placards urging Frazier to "take a shot at a real fighter" as Cloverlay convened a board meeting to weigh an offer by Malitz on May 10, 1967. In April, with Doug Jones behind him, Frazier had fractured the jaw of Jefferson Davis in Miami Beach on his way to a fifth-round technical knockout and then scored his sixteenth victory by winning a ten-round decision over George "Scrap Iron" Johnson in Los Angeles in May. By then, Frazier had ascended to number two in the WBA rankings and had cards to play. Futch claimed he discouraged Durham from entering the tournament, which included opponents who could be a problem for Frazier at that stage of his development—particularly Terrell, Ellis, and Quarry. Shrewdly, Futch told Durham, "Let them fight it out. Why fight through a crowded field when you can wait for the winner?" With no interest in tangling with the potentially dangerous Martin, Durham received the approval of the Cloverlay board to decline the offer by Malitz and instead sign to meet George Chuvalo at Madison Square Garden in July.

With a scarred face that had withstood years of punishment, including fifteen rounds each with Ali and Terrell and twelve with Patterson, Chuvalo had not been knocked off his feet in sixty-two professional bouts. The son of a Croatian-born butcher, Chuvalo grew up in a grim section of Toronto called the Junction, where he learned to box at age nine from lessons he found on cards enclosed in cereal boxes. He won eighteen of nineteen amateur bouts, including the Canadian amateur heavyweight championship in May 1955, and a year later turned pro by knocking out all four of his opponents in the Jack Dempsey Heavyweight Novice Championship at Maple Leaf Gardens. When his career stalled in the early 1960s, he scraped up five thousand dollars to buy back his contract from his manager. Someone had told him that there was a trainer in Detroit who worked well with aggressive fighters—Theodore McWhorter—so Chuvalo packed up his wife and three children and found lodging in the Motor City in a ten-dollar-per-week room, where they subsisted on canned spaghetti and Ritz crackers. "I was living like a dog," Chuvalo told *Sports Illustrated*. With the help of McWhorter and later Irving Ungerman—a Toronto poultry processor who became his manager—Chuvalo acquitted himself nobly. *The Ring* magazine selected his loss to Patterson on February 1, 1965, as the Fight of the Year.

No prior effort in the ring by Frazier highlighted his emerging aura of savagery better than his evisceration of Chuvalo. While he did not "tilt George"—which is to say, become the first to knock him down—Frazier was as unpitying of Chuvalo as he had been of the three sparring partners he had gutted during training camp. He had promised the press that he would "come out smokin'" and he did precisely that. By the end of the first round, Chuvalo was bleeding from two cuts—one on his left forehead from an apparent head butt and the other below his right eye. Blood spurted from the eye cut as Frazier bore in on it in the second and third rounds. By the end of the third round, the big-boned Canadian could not see out of his swollen right eye. He later said, "I looked like a one-eyed cat peeping into a seafood store." Early in the fourth round, Frazier pounded two left hooks into the eye that landed with such cruelty that Chuvalo, his face now covered with blood, was certain that his eyeball had been jarred loose. The pain was unbearable. He turned away from Frazier as the referee stepped in and stopped the fight. Doctors later told Chuvalo that his eye had "fallen through the optic floor" and he had been in danger of losing it.

Frazier had come far in the ten months that had passed since that close call with Bonavena. With Futch on board, he had incorporated the bob-and-weave into his style. He came at his opponents at angles, no longer walking in unprotected, yet critics contended he remained easy to tag, far too willing to sacrifice his body in order to get his shots in. On the other hand, it was widely agreed upon in the wake of his destruction of Chuvalo that he was the hardest hitter in the heavyweight division. Moreover—and the fans picked up on this—he mowed down his opponents with cold finality, as if he were back on the slaughterhouse floor in a bloody apron. Invariably, there would be a smile on his face as he eyed his foe, as if—according to Futch—"he was hungry and in search of something to eat." Chuvalo attested to the ferocity with which Frazier had come at him in an interview with my father for *Sports Illustrated*. In his hotel room on Eighth Avenue the evening after the bout, Chuvalo said Frazier was so fast that it was as if he were hitting him with four hands. "Everything moves, his head, his shoulders, his body and his legs," said Chuvalo, his face swollen and bruised. "Meanwhile he keeps punching and putting on pressure. He fights six minutes every round. . . . Whoever gets him from here will catch hell."

Over at the old Garden, Markson told Brenner, "Teddy, you better start thinking about a big show for the new Garden."

"I got some ideas," Brenner said.

"So have I," Markson replied.

———

For a young man who had climbed out of the cradle of poverty, it was a big moment for Joe when he could afford to buy his own house. He and Florence and their three children had lived with Aunt Evelyn for what had seemed like ages, and in 1966, as his career began to take off with the formation of Cloverlay, he found a row house on Ogontz Avenue, in the West Oak Lane section of Philadelphia. There were four bedrooms—one occupied by daughters Jacquelyn and Weatta, and another by son Marvis—and in the finished basement there was room for them to play. For an urban neighborhood, it was relatively safe, yet not without occasional incidents. The races largely coexisted with equanimity in this small patch of Philadelphia, even if that was not the case in more troubled sections of the city. Weatta would remember that an elderly white woman across the street offered piano lessons. Florence always had soul food simmering on the stove—fried chicken, collard greens, pigtails, and such. When Joe Hand once dropped by, he found what appeared to be an ornament sitting on the lawn. Florence had redone the bathroom and repurposed the old sink as a bird bath. That was how they did things back home; nothing went to waste.

Joe vowed to himself to be the same reliable provider his father Rubin had been. He saw to it that Florence had money to run the house and that the children not only had their needs but their desires met. "Dad was involved in our lives," said Weatta. "Growing up in the South, he was given everything he needed but not what he wanted. So if we wanted something and he was able, we got it." He and Florence accompanied them to services on Sunday at the Bright Hope Baptist Church, and, though Joe himself had not been attentive to schoolwork, he demanded that his children hit the books. Around the house, he always seemed to be tinkering with the air conditioner or movie projector or some other household item that was on the fritz. "Why pay somebody to come in and fix it when I can do it myself?" handyman Joe would say. He always had a roll of electrical tape on him. To relax with his children, he would play table games with them in the evening—blackjack, checkers, or Monopoly. Marvis remembered that he would "play monster" with him and his sisters: "Pop would pretend he was asleep and we would crawl on top of him. He would hold us for a couple of seconds and then let us go." By contrast, he would be far less involved with his children by Rosetta in New York—Renae and Hector. Given the hard feelings that Joe was unable to overcome, there would be years of separation.

Florence would have a third daughter, Jo-Netta, and become pregnant with a fourth, Natasha, while they lived on Ogontz Avenue. With four children and a fifth on the way, and with the arrival of cousin Vernell Williams and her small child, the house had become suddenly too small, so it was not long before Joe and Florence were looking for a larger place. Also, Marvis was harassed by school bullies who knew how his father earned a living and spoiled for a fight. While Marvis held his own in these encounters, Joe presented himself at William Rowen Elementary to discuss the issue with the principal. Joe remembered how it had been down in Beaufort, how he had faced down the same sort of "scamboogahs" in his youth, and he did not want his oldest son to have any part of it. As Ogontz Avenue fell into decline with the encroachment of gang activity, it became clear that a move was in order. By the end of the 1960s, the Fraziers could afford to move to a better school district in a safer area.

No son ever gazed upon a father with more worshipful eyes than Marvis did. He was six when he first saw his father fight— against Chuvalo in that bloodletting at the Garden—and thought of him as invincible. To Marvis, no man loomed larger in his young world. With a laugh years later, Marvis would remember: "When Pop said, 'Jump,' I asked, 'How high?'" On those few occasions when Marvis would step out of line, Joe would shoot him a hard look and say, "Come over here." Joe would then ball up a fist, hold it up for Marvis to inspect, and say, "Smell this." When it came to keeping order in the house, of seeing to it that plates of food were finished and lights were turned off, Joe could get more done with a stern look than his mother, Dolly, could with a quiver of switches. But Joe also heaped affection on his children the same way Rubin had on him, and it was always a sad day when he would leave for training camp, whereupon Marvis would become the man of the house and Joe would become a nightly voice on the telephone.

In the arboreal solitude of his training camp at the Concord Hotel, in the Catskills, Joe pushed himself hard and his sparring partners harder. He was unsparing with them in the ring, just as he had been back at the Twenty-Third PAL. The beatings to which he subjected them were so severe that it was not uncommon for Durham to have to bring in replacements. "He was an animal," said Lester Pelemon, an assistant trainer. "The more he hit you, the more he wanted to hit you. And if you hit *him*? That really pissed him off." Often in the evening, he and his sparring partners shot craps; Moleman Williams once took five thousand dollars off Joe. Whenever Joe dropped a roll—and that was commonplace— Pelemon remembered that he would instruct him to arrange the order of sparring partners the following day so that the big winner would get in the ring with him last. "By then Joe would be warmed up and really get his licks in—you know, in revenge," said Pelemon. Appointed camp "snitch" by Durham, Pelemon remembered that once while Yank was away, Joe wheeled out his Harley-Davidson and informed Pelemon he was taking a spin. When Pelemon re- minded him that Yank had ordered him expressly not to allow Joe on "that damn bike," Frazier replied that *he* was the boss.

"So Joe got on the bike and he was out there showing off and the bike went out from under him," Pelemon told me. "The bike spun in circles and ended up in a ditch. When Joe got up, he was all skinned up and bleeding. He said to me, 'You better not tell Yank!' And I said, 'Tell him? All he has to do is look at you.' Sure enough, when Yank came back and got a look at Joe, he shouted, 'Les, what the hell went on around here?'"

To pass the hours in the evening at camp, Joe ran up colossal phone bills, and not just because he was calling home with regularity. When Cloverlay asked Joe Hand to have a word with him on the subject, Frazier told him: "No problem. These are girls. I'll just drive to the city and see them." Hand replied, "No, you're better off using the phone." Except when he was in camp and observing a pledge of celibacy, Frazier let the good times roll when it came to liaisons with women. One could say without fear of contradiction that he was prolific in his sexual adventures, perhaps not to the standard of NBA goliath Wilt Chamberlain—who claimed in his autobiography that he had had twenty thousand liaisons— but stride for stride with Ali. For his part, Joe looked upon his dalliances as a prerogative due him once the bills were paid and the children were tucked in. The same fevered blood that had run through Rubin ran through Joe, who observed: "Change me, you'd have to go back and change my daddy." From the vantage point of adulthood, Weatta would just shrug and say of her father, "What can I say? He was a rollin' stone."

Philadelphia Inquirer staff writer Jack Lloyd found Frazier in his dressing room at the Latin Casino in 1970 with the emcee, one "Mr. Scratch"; a sparring partner known only as Pete; and a man who called himself "Unc"—the very "Uncle Cadillac" who sister Mazie had been so worried would be a corrupting influence on her brother. From across the room, Lloyd overheard part of a conversation in which Frazier said, "Are you kidding? My wife would raise hell." Frazier explained that Florence did not "think too much of this show business thing," yet added that she knew he was a "good husband." By a "good husband," it seems likely that he was refer-ring to himself as good provider.

"No wife likes it, man," Unc said. "They know the temptation is there."

"Yeah," Frazier said, "but she knows I am a good husband."

Lloyd observed: "Unc shook his head gravely."

———

Scanning the far reaches of the new Madison Square Garden as he stood in press row, Jimmy Cannon turned to Dave Anderson and observed, "You'd have to have an arm like Roberto Clemente to hit us with a bottle from up there." Far more than just a build-ing to stage boxing events—which had been the original intent of the now-decrepit old Garden—the $150 million Madison Square Garden Center housed not just the arena itself but six other facili-ties, including the Felt Forum (a 5,227-seat concert hall); a forty-eight-lane bowling center; the National Art Museum of Sport; a Hall of Fame of Garden heroes; a rotunda for trade and sport shows; and a 486-seat movie theater. To clear space between Seventh and Eighth Avenues and from Thirty-First to Thirty-Third Streets, the Pennsylvania Railroad Station had been demolished, amid fruitless protests by dissenters that included the *New York Times* editorial page, which called it a "monumental act of vandalism against one of the largest and finest landmarks of its age of Roman elegance." Although passengers would still arrive and depart from the lower floors of the new Garden, it was without the grandeur that travel to and from there once inspired. Yale University architecture histo-rian Vincent Scully lamented the passing of that era: "One entered the city like a god; one scuttles in now like a rat."

Bob Hope and Bing Crosby teamed up to open the arena on February 11, 1968, with a gala event billed as "The New Garden Salute to the U.S.O." The comedian and the crooner had not performed together "live and in person" since 1942, yet here they were, the top bananas on a star-studded revue that included Pearl Bailey; Joey Heatherton; Jack Jones and his wife, Jill St. John; Phyllis Diller; and Les Brown and his Band of Renown. In a boxing spoof in which he squared off with unbeaten former heavyweight champion Rocky Marciano, Hope called himself "Chicken Delight" and told the 19,870 patrons in attendance that in the 1920s in Cleveland he appeared in three amateur bouts as "Packy East." He won the first two but said, "The last time, I got knocked right into dancing school." The laughs would cease and the action would become genuine three weeks later when the new Garden jumped in with its first boxing event: Joe Frazier against his old rival Buster Mathis. The winner would be recognized as heavyweight champion in New York, Illinois, Massachusetts, Maine, Central and South America, and Asia, and was then slated to meet the winner of the WBA tournament to unify the title that had been torn from Ali.

As Frazier and Mathis engaged in preparations for their March 4 bout, the eight-man tournament had been whittled down to Jimmy Ellis and Jerry Quarry. At the Houston Astrodome in the quarterfinal round that August, Ellis had stopped Leotis Martin on a technical knockout in the ninth round of a bloody bout in which Martin nearly had his bottom lip sheared off. That October in Los Angeles, Quarry prevailed in his quarterfinal bout by edging Floyd Patterson in a twelve-round majority decision that provoked boos from the crowd at Olympic Auditorium. "You lost, Quarry, you lost and you know it. You got a pass, you bum," yelled a fan as Quarry walked up the aisle to his dressing room. At Freedom Hall in Louisville in December, Ellis scored a unanimous twelve-round decision over Oscar Bonavena, who had advanced to the semifinals by beating Karl Mildenberger in Frankfurt in September. Ellis floored Ringo in the third and again in the tenth round. Quarry then faced Thad Spencer, who had gained his semifinal berth by upsetting 8–5 favorite Ernie Terrell the previous August. With a six-day stubble of beard, Quarry stopped Spencer on a technical knockout with three seconds remaining in the twelfth and final round in Oakland. Quarry had told the *New York Times* prior to the victory: "Boxing needs a white champion to replace Cassius Clay." Ellis and Quarry would meet in April.

To get some work in prior to the Mathis bout, Frazier had been fed two soft touches in the fall of 1967. In the first sports event at the new Spectrum in Philadelphia, Frazier stopped an overmatched Tony Doyle before a crowd of only 8,404 fans. To hype the event, Doyle, of Salt Lake City, Utah, and his manager, Angelo Curley, had circulated the amusing fiction that Doyle had beaten Frazier four years before at the National Amateur Athletic Association tournament in Utica, New York. Stan Hochman exclaimed in the *Philadelphia Daily News:* "Yo, didn't they think someone would look it up? Didn't they think *The Utica Observer* has files?" Frazier told the press in his dressing room that he was looking forward to taking some time off. He explained to *Philadelphia Daily News* columnist Tom Cushman: "I've been away from my family too much this year." But that vacation would not begin until he traveled to Boston and faced former New England heavyweight champion Marion Conner in December. With a thirty-one-pound weight advantage, Frazier stopped Conner on a technical knockout in the third round and for his effort picked up five thousand dollars. Next up: Buster Mathis, "The Big Bus."

In the three and a half years that had elapsed since he had faced Frazier as an amateur, Mathis had shed seventy-five pounds and now tipped the scales at 240. Early on, he was conditioned by the eccentric trainer Cus D'Amato, who delivered on his promise that he would work Buster until his blisters bled. By virtue of cutting calories and hours upon hours of roadwork, the weight came off. But while Cus was an extraordinarily fine teacher, Buster found him to be so disagreeable that his management team fired him. The group—Peers Management—was a consortium of young bluebloods headed by Jimmy Iselin, whose father was president of Monmouth Park racetrack and a director of the New York Jets. With an eye toward bagging that rarest of athletic prizes—a heavyweight champion—Iselin and his cohorts had looked far and wide for the perfect candidate. Of the young heavyweights they scouted, including Frazier and Quarry, Iselin claimed that Mathis graded highest in personality, speed, size, and strength. Peers invested sixty thousand dollars to get Mathis up and running, part of which included the construction of a gym in Rhinebeck, New York. "We brought him along slowly, the way you would a racehorse," Iselin said. To that end, Cus lined up a string of pushovers for Buster, just as he had done with Floyd Patterson years before and would do with Mike Tyson years later. Mathis came into the Frazier fight with an unblemished record in twenty-three pro appearances, yet he remained unranked until *The Ring* magazine slipped him in at the bottom of the Top Ten only days before the bout. He had not been invited to participate in the WBA tournament.

Ranked number one by *The Ring* but only number eight by the WBA—he had been dropped from number two when he declined to appear in their tournament—Frazier commanded $175,000 for the Mathis bout, one hundred thousand dollars more than Buster received and fifty thousand more than he would have been guaranteed by Malitz if he had advanced to the finals. In order to ensure a lucrative gate that would enable them to pay Frazier his price, Markson and Brenner sweetened the card by adding a championship match between middleweights Emile Griffith and Nino Benvenuti. At the Concord Hotel in the weeks leading up to the bout, Frazier forged a friendship with Griffith. They sparred together in a closed-door session, offered each other tips, and in the evening could be found shooting craps. Nearing the end of camp, Joe called down to Bright Hope Baptist Church on a Sunday and spoke to the congregation over loudspeakers. He told them, "Keep praying for me." But to hear his banter with reporters, it was Buster—and not he—who needed to call upon divine intervention. "He runs like a thief," said Frazier, who was still irritated by his two losses to Mathis as an amateur. From across the Hudson River in Rhinebeck, Buster shrugged and called Frazier a "nitwit." Down in Philadelphia, Cloverlay sold close to eight hundred tickets among its shareholders and chartered a special eight-car train for the group to travel to Manhattan. Holders of ringside seats were asked to wear formal attire.

Some three hundred so-called black militants armed with placards paraded up and down the sidewalk outside the Garden in protest of the unceremonious dethroning of Ali. By and large a peaceful assembly, it underscored the widely held contention that, as Sandy Grady observed in the *Philadelphia Evening Bulletin*, the pairing of Frazier and Mathis to determine a champion was "a flashy hollow charade as long as Clay is still the best fist fighter in the world." But the evening proved to be just the blockbuster that the Garden had hoped it would be, as a crowd of 18,096 fans produced a record indoor gate of $658,503. On the heels of the Griffith-Benvenuti bout, in which the handsome Italian won the title for the second time in a unanimous fifteen-round decision that Red Smith dismissed as a "dull affair," the showdown between the big boys was nothing if not entertaining. With a thirty-nine-pound weight advantage and that three-and-a-half-inch edge in reach, the six-foot-three Buster appeared to tower over Joe, who not only was three and a half inches shorter but boxed out of a crouch. Through the early rounds, Buster was up on his toes, holding Frazier at bay with what closed-circuit announcer Don Dunphy described as "catlike moves." But Frazier bobbed and weaved as he poured on the pressure and adhered to the strategy that Durham had articulated to him: keep working that body until his hands drop, then go for the head. Near the end of the second round, Mathis began bleeding from the nose. Early in the third round, the pace was so intense that Dunphy observed: "Something has got to give."

Beginning in the fifth round, it did. While Mathis appeared

to have been in control prior to that, Frazier began to take command as his big target covered up in the peekaboo style that had been impressed upon him by D'Amato. Dunphy spotted the shift: "Frazier seems to be cutting him down little by little." Though Mathis had a good round in the sixth, during which he counterpunched effectively and scored heavily, he continued to wear down as Frazier dug in, grunted, and pummeled him with left hooks. To slow Joe down—and catch a breather—Mathis grabbed and held on, which inspired Grady to observe in the *Bulletin:* "They say Buster is built like an NFL lineman. Sure, and last night he would have drawn 1,000 yards in holding penalties." By the tenth round, the crisp white trunks that Mathis had worn in the ring were crimson from that bloody nose, which had only worsened as the bout had progressed. The big man no longer had any spring in his legs. Of the uninterrupted beating Mathis was now absorbing, Dunphy observed: "Frazier is just winding up and sending them in." To get him to reengage, Frazier urged Mathis: "Come on!" With five rounds remaining in the scheduled fifteen-round bout, Mathis found his corner at the end of the tenth and lowered his battered body upon his stool.

"How far am I behind?" he asked Fariello, his trainer. "Can I pull it out if I win the last five?"

"Easy," Fariello told him. "But stop laying on him, Buster. Keep him off!"

Frazier sat in the opposite corner with the easy manner of a man on a coffee break.

"What round was that, the tenth?"

Durham hovered over him and teased, "Naw, the fifth."

"Darn," Frazier said.

"Keep going at that body," Durham told him. "If you see his hands drop, take your shot."

Think of a drill press puncturing a piece of sheet metal—that was how suddenly the left hook by Frazier in the eleventh round dropped Mathis. Stunned, he swayed as if blown over by a hard wind and fell to the canvas, his body splayed across the bottom strand of the velvet ring ropes. By the count of nine, he had grabbed the top strand and pulled himself up. But referee Arthur Mercante stopped the fight, officially at 2:33 of the eleventh round. Frazier jumped in the air and was embraced by Durham. With his handlers on either side, Mathis began the slow walk back to his dressing room, where he sat for an hour before inviting in the press. Again and again, he said he could not believe "it was me fighting so poorly." His eyes glistened with tears. Iselin agreed that it had been a "poor showing" and conceded, "Maybe you writers were right. Maybe we should have fought better opponents. . . . But he was so sure he would beat Frazier." Iselin added that Buster would take a week or so off and "think about whether he wants to stay in boxing." Over in his dressing room, Frazier found himself running out of words to express his joy. While he had found Mathis to be troublesome in a three-round amateur bout, he had been certain that he could wear him down over a longer distance, particularly given the manner in which Mathis had engaged him—"stoopin' down under me, all covered up like a turtle. All I had to do was go bam, bam, with hooks and uppercuts."

Cloverlay held a victory party at the Iron Horse Restaurant, beneath the Garden in Penn Station. With the homeward-bound train not scheduled to leave until 2 A.M., the cocktails flowed amid a haze of cigarette smoke and speculation of even greater days ahead. Who would Joe fight next? Quarry? Ellis? Or would it be Manuel Ramos? He had been talked up in the papers. Whoever it would be, no one doubted in that heady moment that Joe could lick anyone—including Cassius Clay, or Muhammad Ali, or whatever he called himself. Wearing a red beret with a green towel draped over the shoulder of his overcoat, Frazier was greeted by raucous applause when he stopped by to engage his backers. But he did not stay there long. With Florence on her way back to Philadelphia—

she had come to New York but had not attended the bout—Frazier returned to the City Squire Motor Inn, at Seventh Avenue and Fifty-Second Street, where there was a big party that only began to wind down at dawn. Frazier opened the door to his suite and found snoring bodies plopped on the sofa, chairs, and floor. In the bedroom he found two young women stretched out on the bed, one with her head at the bottom and the other at the top.

The young women were Denise Menz and her roommate, whose parents had a connection to Cloverlay. When her roommate had invited her to join the three of them for a night out at the fights, Denise enthusiastically tagged along, first stopping off for prefight cocktails at a Cloverlay gathering at the City Squire. It was there that she learned that there was an unexpected shortage of two tickets, so Denise and her roommate stayed behind in the Cloverlay suite and later joined in the victory party. "We were up all night dancing, pretending we were movie stars," said Denise, then nineteen. Not a sports fan, she had no idea who Joe was when he stopped in the bedroom and engaged them in small talk. When he got up and said, "Well, I gotta go. I have things to do," she presumed she would never see him again. But ten seconds later he tapped on the door and sat down with them to talk some more. And Joe told them, "Next time, I'll make sure you get tickets."

CHAPTER FIVE

SKY LARKING

Joe, peeking inside an envelope at the proceeds from his victory over Manuel Ramos, with Gypsy Joe Harris (L) and Emile Griffith (R), 1968.
Philadelphia Bulletin

Whatever measure of personal accomplishment Joe Frazier gleaned from beating Buster Mathis in March 1968, it was hard not to look upon his small share of the heavyweight championship as scarcely more than an attractive piece of costume jewelry—appealing to the casual eye, perhaps, yet of dubious genuine value given the unjust exile of Muhammad Ali. Seven weeks later, there would be a second petitioner for the vacated crown with the emergence of Jimmy Ellis, the winner of a fifteen-round decision over Jerry Quarry in the finals of the WBA tournament in Oakland. As the United States was drawn ever deeper into the Vietnam War and widespread violence erupted in the wake of the assassinations that spring of Martin Luther King and Robert F. Kennedy, Ali remained free on bail pending the appeal of his case to the Supreme Court, yet he was prevented by the governing bodies of boxing from pursuing his chosen livelihood. To keep body and soul together, he took speaking engagements on college campuses for a few thousand dollars a throw. At an appearance at Princeton University, where he spoke before a standing-room-only crowd of one thousand on a variety of subjects that included his obedience to Allah and his allegiance to the Nation of Islam concept of racial separatism, Ali called himself "the fastest and the best boxer alive today" and poked fun at Frazier with a fragment of verse warmed over from the Liston days.

> *Ali hit him with a hard right.*
> *"Oh man, what a beautiful swing!*
> *"Then a left and right, and the champion*
> *"Punches Frazier right out of the ring!"*

Walled in by the inescapable shadow of Ali, Joe first defended his portion of the heavyweight championship against Manuel Ramos, of Mexico City, on June 24, 1968, at Madison Square Garden. Only 6-6-2 in his first fourteen bouts as a pro, due in part to an undisclosed fracture to his right hand, Ramos had since strung together fifteen consecutive victories, including twelve by knockout. Notably, he defeated Eddie Machen and Ernie Terrell during that stretch. He was a good puncher with a solid chin. Unlike the "runners" to whom Frazier had become accustomed, Ramos figured to stand in and slug it out. Proudly, he reigned as the heavyweight champion of Mexico, which Sandy Grady of the *Philadelphia Evening Bulletin* compared to "being the ski jump champ of Dry Gulch, TX." Stan Hochman of the rival *Daily News* chimed in:

"He looks like Ricardo Montalban applying for the job of chauffeur in an old Esther Williams movie." Giving away four inches in height, four and a half pounds in weight, and two and a half inches in reach to the six-foot-three Ramos, Frazier emerged from five weeks of tearing apart his sparring partners at the Concord Hotel at a trimmed-down 203½ pounds. Las Vegas installed him as a 4–1 favorite. While Ramos was no pushover, only a noisy contingent of countrymen decked out in floppy sombreros gave him any shot at an upset.

Charging from his corner at the opening bell, Ramos stood toe-to-toe with Frazier as they exchanged ordnance. Early in the first round, Ramos caught Joe with a solid right hand to the jaw that buckled his knees and rocked him back on his heels. The action was so intense that veteran broadcaster Don Dunphy exclaimed from ringside: "With the exception of probably [Jack] Dempsey and [Luis] Firpo, this is the greatest first round of a heavyweight championship fight that I can remember" (an opinion later backed up by Nat Fleischer, the ancient editor of *The Ring*). With the round nearing its end, Frazier pinned Ramos in the corner with a barrage of head and body shots and staggered him with a left hook. Only the ring ropes prevented him from falling. Urged between rounds by Yank Durham to "get closer, cocksucker—and stay down lower," Frazier sprang out of his corner for the second round and floored Ramos for the first time in his career with a combination right uppercut and left hook. Ramos climbed to his feet at the count of nine. From that juncture, Frazier extracted any resolve that remained in his dazed opponent with a fierce and sustained body attack. With just seconds remaining in the round, Frazier slammed Ramos with a left hook that snapped his head back and sent him spilling to the canvas. Although Ramos was up at the count of two, his glazed eyes betrayed the unambiguous plea of surrender. Referee Arthur Mercante waved the bout over.

Cheered for his gallantry as he departed the ring, Ramos told the press, with the help of an interpreter, "I am ashamed." According to *Philadelphia Inquirer* columnist Sandy Padwe, Ramos "broke down and sobbed" later, in the privacy of his dressing room. Over in his own dressing room, Joe said he was glad it ended when it did, saying: "Why should he get hurt?" He liked Ramos. While Frazier conceded that he had been stunned by Ramos in the first round, he had quickly pulled himself together and launched a counterattack instead of backing up, which would have been contrary to his burn-down-the-village style. Frazier explained, "If I run and hide, Ramos will turn killer. I had to retaliate." Financially, the Garden took a loss on the event, yet Markson and Brenner shrugged it off as goodwill in the expectation of doing bigger deals with Cloverlay. Slipping out of his boxing gear and into a blue suit accented by blue alligator shoes, Frazier said that unless the unforeseen happened and "Clay" was sprung loose from his legal woes, he planned to take the balance of the year off to give his singing career more attention.

Far more than just a lark to keep him busy between bouts, Frazier approached show business with the vigor of any accomplished professional. While his voice would never be as remunerative as his fists or bring him even close to the same level of critical acclaim—quite the opposite, actually, in that it was expensive and the reviews tended to the sour side—he gravitated to the stage when time allowed. In the same way he had applied himself to boxing back at the Twenty-Third PAL, day by day shedding that excess weight in a quest for some larger version of himself, he was certain that through a commitment to hard work he could accomplish similar gains as a vocalist. Although there would be problems at home because of the time it ate up, the late nights, and the way of life that merged with his increasing wanderlust, he found in song an avenue to express himself in a way he could never do in words, and surely not with gloved hands. Even if his bookings had less to do with his virtuosity as a performer than with his drawing power as a champion prizefighter, no one could say that he did not give his all.

The occasional split trousers he would suddenly find himself with onstage attested to that.

Joe had embraced music since childhood. Along with singing hymns in the church each Sunday in Beaufort, he would join in with the quartets that formed in the front yard on summer evenings. Not unlike the way teenagers would harmonize on street corners in South Philadelphia, the a cappella groups Joe joined in with in Laurel Bay performed spirituals, or they would extemporize in what sister Mazie called "a kind of rap." When instruments were called for, they clapped their hands and drummed on the bottom of a tub. "It was what we did for entertainment," Mazie said. "When we came in from the fields at the end of a long day, it was the way we had of relaxing. In those early days, we had no TV." By wide agreement, Joe had none of the vocal skills possessed by his older brother, Rubin Jr.—also called Jake—a church deacon who sang in the choir and years later with a group he helped found, the Gullah Kinfolk. But Joe sang then and later for the sheer joy of it, especially in the car on those long drives he preferred to take instead of flying. As the eight-track player in his Cadillac blared the stylings of the soul stars of the era—James Brown, Sam Cooke, Bobby Womack, and scores of others—Joe would sing along, often the same song again and again. Denise Menz used to sit in the passenger seat and bury her head in a pillow. Lester Pelemon remembered Joe had a fondness for "Proud Mary" by Creedence Clearwater Revival, "Knock on Wood" by Eddie Floyd, and "Mustang Sally" by Mack Rice and later Wilson Pickett.

"Joe used to call me 'Puff,'" said Pelemon, who sang with a group called Soul Brothers Six before later joining Joe's group, the Knockouts. "He was 'Smoke' and I was 'Puff.' 'Smoke' would be behind the wheel, start singing 'Proud Mary,' and go, 'Come on, Puff, jump on in.' And I would think, 'Not again!' But he loved that song, and I sang along with him as we drove down the highway."

Originally, Joe formed the Knockouts with a group of friends, some of whom he had grown up with in South Carolina and who had found their way to Philadelphia. One of them was Bobby Kears, who played the bass guitar and for whom Joe purchased a nine-thousand-dollar dialysis machine a few years later when he developed kidney problems. "We had heard Joe wanted to form a band," Kears told Hochman in the *Daily News*. "One night we all got together. . . . Joe liked the way we played." Rehearsals were held on the second floor of the Twenty-Third PAL, once Joe had finished working out for the day. In April 1968, Cloverlay arranged for him to cut two singles on their label for local release: "Come and Get Me, Love" and "The Bigger They Come, the Harder They Fall." Jerry Gaghan dropped an item in his showbiz column in the *Philadelphia Daily News* that speculated Cloverlay would be "taking a big batch of the biscuits." With the release of two more singles that summer aimed at national release—"You Got the Love" and "Good News"—he and the Knockouts headed down to Atlantic City in early August to play in front of a live crowd.

The place was called the Jet Set Bar and Lounge. Standing at the door was Ben Anderson, a vice-squad cop who was said to own a piece of the operation under the table. Joe was up onstage in a tuxedo singing when there was a noisy commotion in the lobby, where Ali appeared with a hundred or so followers. They were soaked. Along with a companion, Ali had braved a downpour to hand out Nation of Islam literature in a stroll along the North Side, the crowd behind him growing with each soggy step. Anderson stopped him at the entrance of his club, whereupon Ali spotted Frazier onstage and began heckling him amid the laughter of the audience. For someone who looked upon his singing career as a serious undertaking—and Joe did—this unexpected visit by Ali had to be an irritating intrusion. But Anderson let Ali pass and Frazier played along as the dethroned champion joined him onstage. "I'm going to jail," Ali announced. "I'd rather be in jail than in Vietnam dead." At that point, he and Frazier stripped out of their jackets and began sparring, as Ali dazzled the room with his incomparable footwork. "This is my man," Ali said of Frazier. Then, just as suddenly as he had appeared, he stole away back into the rain.

Joe and the Knockouts debuted formally at the fourteenth annual Hero Scholarship Thrill Show at JFK Stadium in Philadelphia on September 7, 1968, before a crowd of ninety thousand. Wearing a fire-engine-red jacket and shoes, black trousers, and a white turtleneck, Joe hopped onstage as if he were bounding into the ring and sang "You Got the Love" and "The Bigger They Come, the Harder They Fall." On hand that day was local disc jockey Jerry Blavat—"The Geator with the Heater"—who enjoyed a bond with the black community because he played the latest from Motown instead of the Beatles and featured acts such as Sly and the Family Stone, Peaches and Herb, and the Temptations on his weekday television show on WFIL, *Jerry's Place*. Blavat became fond of Joe and invited him to become a regular on the show, yet when asked to assess the level of skill Joe possessed as a performer, he paused and with a shrug observed: "It was a good thing he was the Champ."

Frazier was asked to appear two days later on the seventeen-hour-long WFIL-TV Variety Club telethon on behalf of handicapped children, hosted by comedian Joey Bishop. Among the lineup of stars were Gladys Knight and the Pips, Leslie Uggams, Bobby Rydell, Rodney Dangerfield—and Muhammad Ali. A former South Philadelphian who had come to prominence alongside Frank Sinatra, Dean Martin, and Sammy Davis Jr. as a member of the Rat Pack, Joey sparred with Frazier onstage, even at one point accidentally grazing his nose with a knuckle. Ali got in on the act later in what Hochman described in the *Daily News* as a "wild charade, the kind of thing that used to make people laugh, and now only leaves them squirming." Instead of "The Ali Shuffle," he did "The Uncle Tom Shuffle." Ali slipped off the diamond pinky ring Frazier was wearing and tried it on. Playfully, he began throwing jabs at Frazier, who slapped them away with what Hochman called a look of "churning puzzlement." On the heels of what had happened in Atlantic City, Frazier had come to his breaking point. According to Hand and Pelemon, Frazier was helping to unload equipment from the trunk of his Cadillac outside the Academy of Music that Sunday morning when Ali turned the corner with a parade of fifty or so fans behind him.

Ali shouted: "Joe Frazier! Joe Frazier! Joe Frazier!"

Frazier jerked his head over his shoulder. Pelemon told me there was fury in his eyes, adding: "Joe was sweating bullets."

Frazier then reached into the trunk for a tire iron and growled, "I'm gonna put an end to this sucker right now."

———

A photograph once appeared in the *Philadelphia Evening Bulletin* of Yank Durham, clownishly posed in boxing trunks, with Joe Frazier and Gypsy Joe Harris on either side. The two Joes were the talk of Philadelphia in 1968, and Durham had a piece of them both. Gypsy Joe had been on the cover of *Sports Illustrated* the year before yet would come to harbor a sibling rivalry with Frazier that was laced with jealousy. With the capital of Cloverlay behind him and the certainty of some big paydays ahead, Frazier had the undivided attention of Durham in a way Gypsy never would. While Gypsy Joe had no one but himself to blame for his long fall, it would always seem strange to him how the Pennsylvania Athletic Commission suddenly "discovered" his blind eye later in 1968 and proceeded to take his boxing license. He told Robert Seltzer of the *Philadelphia Inquirer*: "Let me ask you a question: How can a man have a license to examine a person and not realize I was blind? I turned professional in '65 and they stopped me in '68. Were the doctors sleeping from '65 to '68?" Given his special appeal, had it been "fixed" so he could box, only to later have it "unfixed" so he could not? But why? Whatever factors came to bear, Gypsy Joe told my father in *Sports Illustrated* in March 1969: "In one second I was dead."

Intrigued by the scope of the talent Gypsy Joe possessed, Durham once attached a custodian to shadow Gypsy Joe for two hundred dollars a week and get him to the gym each day. The bodyguard came back and told Durham: "Save ya money, Yank. Gypsy, he leaves tracks Tonto can't follow." With a cap atop his shaved head and a pocketful of candy in his leather jacket, he would drop out of training on a whim and disappear for days. Once asked where Gypsy Joe could be found, middleweight Bennie Briscoe shrugged and replied, "Bendin' over a pool table, I suppose." Frazier took Harris under his wing, invited him to dinner, and even slipped him occasional "love" from his sock, yet Gypsy Joe could not be persuaded to curb his profligate ways. George James, his trainer, shared an apartment with Harris and remembered how, hours prior to a rematch at the Arena against Miguel Barreto, co-manager Willie Reddish handed him fifty dollars and said, "You know where Joe is. Come on now, go get him for me." James found his roommate in a bar. "I took him to the Arena drunk as hell," said James. Gypsy Joe flopped on a cot in the dressing room and began snoring. As the bout approached, Durham glanced at James and said, "Wake that son of a bitch up, George." James dampened a towel with cold water, jarred him awake by wiping him down with it, and told him, "Put your fucking shit on." James remembered, "So he got into his gear and dressed that Mexican boy for ten rounds like nothing had ever happened."

Notwithstanding his dreadful diet, untimely boozing, and erratic gym attendance, Gypsy Joe had clicked off twenty-four consecutive victories by the summer of 1968, including a nontitle bout victory over welterweight champion Curtis Cokes. But he had become an increasing headache for local promoter Herman Taylor, an octogenarian who did the legwork for Tex Rickard in setting up the Jack Dempsey–Gene Tunney bout at Municipal Stadium in 1926. For the Barreto fiasco, Gypsy Joe had weighed in ten and a half pounds over the welterweight limit and was suspended by

Pennsylvania Athletic Commission Chairman Frank Wildman for sixty days for "jeopardizing the show." Still a believer—yet an increasingly agitated one—Taylor signed him for a March bout at the Arena against former number-one welterweight contender Manny Gonzalez, only to have that postponed once and later scrapped at considerable expense when Harris came down with "acute laryngitis and a respiratory infection." When Gypsy Joe finally fought again, it was as a middleweight, on August 6, against savvy veteran Emile Griffith at the Spectrum. Again, Gypsy Joe showed up overweight at the noon weigh-in and was forced to shed two and a half pounds at the gym in order to come in under the 160-pound middleweight limit. Did that explain his obvious sluggishness in the later rounds of the bout? Whatever accounted for it, Griffith spirited away a unanimous twelve-round decision before a then-Pennsylvania-record indoor crowd of 13,875. The loss cost Gypsy Joe a shot at champion Nino Benvenuti.

Gypsy Joe earned $12,500 for the Griffith bout. "I held it for him in the trunk of my car," James told me. "A week or so would go by and he would be broke and I would go get him some." As always, he was a free spender when he had money, stopping youngsters up and down Columbia Avenue and asking how old they were. If they said they were under twelve, James said Gypsy would slip them a fifty- or even a hundred-dollar bill. On his excursions through North Philadelphia, passersby would spot him and call out to him, "Hey, Gypsy Joe! When you fighting again?" And he would reply, "Keep an eye out for those posters." They were always stapled to utility poles. As Taylor set the date for the Gonzalez bout—October 14—the word out of the gym was that Gypsy Joe was a new man on the heels of the Griffith setback; even Gypsy Joe said as much in the paper. But quite another report circulated back to Wildman, who had heard that Harris had not been working out. Wildman called him in for a hearing, pointed to Taylor, and said: "There is no reason why this man has to sweat and worry until

October 14. You owe it to him and you owe it to the fans to come in at the specified weight." With the bout still a week away, Gypsy Joe had to trim down by six pounds in order to come in at the 156-pound junior middleweight limit or Wildman told him that he would suspend his license again. Whatever room Gypsy Joe still had to heed that warning was foreclosed upon by a heated exchange that James said occurred in private between his fighter and Taylor.

"Joe, you have to stop coming in overweight, you hear?" Taylor told him.

"Leave me the fuck alone," Gypsy Joe shot back.

"What did you say?"

"I told you, man. Leave me the fuck alone." Gypsy Joe then leaned back on the bench that he and James were sitting on and closed his eyes.

James sighed years later. "Next thing you know, the doctors came out and said to him: 'Can we see you in here?' And when he went back into the room, they went right for that eye. And that was the end. I begged him. I begged him, 'Joe, go apologize.' He said, 'Fuck that.' I said, 'Joe, listen to me. You gotta hear me, son. You gotta do it.' Sadly, he never did."

With Gypsy Joe suddenly jettisoned from the scene, Frazier had what appeared to be unobstructed access to the favor of the hometown fans. Nine of his twenty-one bouts had been held in Philadelphia, yet none had stirred much more than passing interest. With the exception of the surprise knockdown by Mike Bruce in 1965, he had breezed through his early undercard fights there. Given the portfolio of some of his opponents, who appeared as if they had been found nodding off on a bench at the bus depot, it had shocked no one when Abe Davis climbed into the ring with a hole in the bottom of a shoe. Others showed up well into the downside of their careers. Four years before Frazier stopped Billy Daniels in the sixth round amid a cacophony of boos, Daniels had given Clay some trouble in their 1962 bout, twice stunning him with right hands before he was stopped on cuts in the seventh. Veteran Doug Jones had far more than just a loss to Clay on his record—he had given him a handful at the Garden in 1963 in a near upset—but he was only a ghost of his former self four years later when Frazier poleaxed him with that left hook to the jaw in the sixth round. Even the tall tale that Tony Doyle had whipped Joe as an amateur was not enough to draw a crowd for the opening of the Spectrum; the seventeen-thousand-seat building was only half full. Observed former *Philadelphia Evening Bulletin* reporter J Russell Peltz, who was only then beginning his career as a boxing promoter: "They ruined Frazier in Philly by putting him in terrible fights."

That did not stop Lou Lucchese from stepping up to the plate. Lucchese owned a toy store in Reading, Pennsylvania, and had a hand in helping to launch Frazier by promoting some of his early appearances. Durham had a fondness for him and once vowed that he would reward his loyalty by keeping him involved as his exclusive promoter. Apparently—or so the tale is told—they were driving back from Pittsburgh after Frazier had beaten Don "Toro" Smith and Durham had been drinking. The promise was quickly forgotten when Brenner later showed up on behalf of the Garden with a bag full of loot. Lucchese was a small operator and was in no position to go big time, yet he had been told by Durham in a more sober moment that he would agree to a defense against Oscar Bonavena at the Spectrum if Lucchese could work out the particulars. Lucchese hopped a plane to Buenos Aires, where Bonavena owned twenty-three apartment houses, two haberdashery shops, a fifty-foot yacht, and a twin-engine plane. When it came down to talking dollars and cents, Lucchese offered him sixty thousand dollars. Bonavena countered at a hundred thousand. They settled at seventy-five thousand, with Lucchese picking up any U.S. taxes Oscar owed. Frazier was guaranteed one hundred thousand dollars, with a percentage of the gate and television receipts.

Like the first encounter between Joe and Oscar, their second one should have been held in a barroom instead of a boxing ring. There was a referee—Joe Sweeney—yet he seemed to be in what an annoyed Durham later called "a fog." Sweeney was unable to hear the bell at the end of each round and permitted the action to continue unpoliced. Early on, Frazier pummeled Bonavena with a fierce body attack, hammering him with slashing uppercuts and hooks. Bonavena covered up in the peekaboo defense then favored by Floyd Patterson as he leaned on the ropes. Red Smith quipped in the *New York Times* that "the unbarbered Argentinian . . . hung there like something in the hall closet." While Frazier was unable to knock Oscar off his feet—including in the fourth round, when he tagged Oscar with twelve consecutive unanswered body blows—he disfigured Bonavena round by round until his eyes appeared as if they were caked with black and blue candle wax. Sweeney took away the eighth round from Bonavena for hitting below the belt, but Durham would later say that "he should have taken away at least three or four more." Correctly, the frazzled referee pointed out that Bonavena and Frazier were both guilty of throwing low blows and asked, "How would it have been possible [to disqualify either] with both fouling?" Frazier won easily on points.

In the crowded corridor outside the dressing rooms, Ali declared that he was unimpressed by what he had seen of the two brawlers. He dismissed it as nothing more than a "slugging match." As he attempted to slip inside to visit Joe, a security guard blocked him and asked to see his press pass. Ali shrugged and said, "All right. I just wanted to wish Frazier luck and tell him I'm going to whup up on him." Ali turned away and announced into a microphone that a reporter held up: "I'm giving Joe Frazier until high noon tomorrow!"

On a table inside his dressing room, Frazier was stretched out in his robe. Someone shouted over the noisy crowd, "Give me a towel! Let me wipe him down!" Nearby was ten-year-old Joe Hand Jr., who had sat with Marvis during the bout and found himself shuttled back into the dressing room. Craning his head above a row of the wagging pens in search of an unobstructed view, he looked on as Frazier was rubbed down by two of his seconds. One worked on his thighs and calves. The other worked on his ears. As Joe Jr. quietly wondered why such careful attention was being paid to his ears, someone called out: "Here comes Bonavena! Give him room." And in walked Oscar, his robe draped upon his shoulders. Both of his eyes were swollen to the size of coin slots. The Argentinian congratulated Frazier and apologized for the low blows.

"Joe stood up when Oscar came in," said Hand. "And when he did, the big jock protector he was wearing fell to the floor. Someone had loosened it so he could breathe better, I guess. But what I remember is how all of this water and blood spilled to the floor. I thought, 'Good God, is he peeing blood?' Bonavena could not see out of either eye. Someone had to hold up his arm so Joe could shake his hand."

Hand paused and added, "I asked Dad on the drive home why they were rubbing his ears. He said they did that to prevent them from becoming cauliflowered."

It was not a good night for Lucchese. Going in, he had predicted a sellout crowd, which would have come to $450,000 in box office. He had scaled his ticket prices accordingly, with ringside fixed at fifty dollars. By fight night, reality had set in. As he scanned the half-vacant building, it became clear he would be hard-pressed to break even on his $250,000 outlay. To do even that, the crowd count would have to come in north of eighty-five hundred. But Lucchese fell short of that figure by two thousand seats, which placed the gate at just over $115,000—less than the $118,000 in box office that the Griffith-Harris bout generated. Crestfallen, Lucchese conceded that he was "disappointed in the Philadelphia fans." But he had an even deeper problem with which to contend. According to Joe Hand Sr., Lucchese did not bring in enough to cover his financial obligations to Cloverlay. "Joe got what he was supposed to get but Lou was still short," said Hand Sr. Cloverlay held a board meeting to assess their options. One was to sue Lucchese and pursue his assets. Hand stood up and asked the board, "Do we want to be known as an organization that shuts down toy stores?" Years later, he observed, "It wouldn't have been in the spirit of what we were trying to do with Cloverlay. So we just dropped it."

But the poor attendance and the lack of support that it signaled weighed on Frazier. By the following May it had become a topic of conversation as he prepared for his bout in Houston against Dave Zyglewicz, whom Frazier pulverized at 1:36 of the first round. In a story by Tom Cushman in the *Philadelphia Daily News* two days before the bout, he said he planned to move from Philadelphia. "You fight in New York, you look down at ringside and see Ed Sullivan, the mayor maybe, big people everywhere," Frazier said. "In Philadelphia, nothing." As an example of the lack of appreciation Philadelphia had for its top athletes—he did not say "black athletes," but that appeared to be the insinuation—he pointed to the vilifications endured by Phillies counterculture star Dick Allen, a would-be singer himself who recorded the single "Echoes of November." As Frazier had done, Allen had played the local clubs with Blavat, who once received an aggravated phone call from Phillies manager Gene Mauch: "What are you doing keeping my ballplayer up all night?" Frazier conceded that he was not yet certain where he was going—only that he was going. And soon. "That you can be sure of," he said, "and the sooner the better." He quipped that he would be treated better in Mississippi.

———

Three weeks before he was scheduled to face Jerry Quarry at Madison Square Garden on June 23, 1969, Joe revealed a piece of himself to the public that had remained hidden. In the lobby of the

Concord Hotel, once again the site of training headquarters, he slammed "Clay" in a conversation with Durham and John F. X. Condon, then in charge of publicity at the Garden. Curtly, Frazier assailed "Clay" as a "big-mouth phony" and a "disgrace." Upon receiving permission from Frazier and Durham, Condon sent a release to the Associated Press. When a reporter followed up, Frazier not only confirmed his comments but also expanded upon them. "What kind of man is this who don't want to fight for his country?" Frazier said. "If he was in Russia, or someplace else, they'd put him up against the wall." On and on, he vented that the public was "fed up" with "Clay" and his "fussin' and fumin'" and that he looked forward to the day when he could "button his big mouth once and for all, knock him out and get rid of him." Frazier added, "I just hope they turn him loose so I can get at him."

It all seemed so unlike genial Joe. Particularly odd was his vociferous condemnation of "Clay" for his refusal to enter the army. Frazier himself had been issued a head-of-household deferment and had stood by Ali not just in word but in deed, using his influence in whatever way he could to help him get his license back. Close observers wondered if there was not some contrived piece of press agentry behind his unexpected outburst, which Condon denied. He swore that Joe actually said what he did, that Durham had spoken of "Clay" in passing and that "Joe just started to take off on him." When Condon told Frazier he would like to use it in a press release, Joe replied, "Go ahead, tell everybody." Cloverlay publicity man Joey Goldstein also disavowed any complicity. But perhaps no one would have questioned the authenticity of his out-of-the-blue comments had it been widely known that less than a year before Joe had come close to wielding a tire iron at Ali. "Unless I had stopped him, the Fight of the Century would have happened right there on the pavement," said Joe Hand Sr. The aborted assault gave a glimpse into the effect Ali had on Frazier, how he not only irritated and distracted him but drove him to the very edge.

Swept up in the whirlwind created by his public attack on "Clay," Frazier remained in a sour mood in the weeks leading up to his encounter with Quarry. The press converged on his camp and found him strangely inhospitable. The *Philadelphia Inquirer* carried a piece with the headline, WHAT HAPPENED TO THE OLD JOE? He bitched at a crew from CBS. *Those lights are too hot. Move 'em back.* Even reporters who had followed him from the very beginning were denied interviews. *Not today. Maybe tomorrow.* When he did talk, there was a sharp edge to his voice, as if he could not get the ordeal over soon enough. The "old Joe" had set aside his disappointment over his loss to Mathis at the 1964 Olympic boxing trials and had the courtesy to stop reporter Jack Fried of the *Philadelphia Evening Bulletin*, shake his hand, and thank him for his help. Five years later, he had a piece of the heavyweight championship; he had flashy threads and wheels and plenty of "love" in the bank (and in his sock); and yet he wore the frown of someone who had just had his whitewalls slashed. Was it the "boredom" of yet another training camp (which was another way of saying he had not had sex for six weeks)? Or had he been spoiled by the success he had experienced so far? Durham blamed the hot weather.

Jerry Quarry could have been a character out of the Leonard Gardner novel *Fat City*. Jack Quarry, his father, had come out of the East Texas dust bowl, the son of an escaped convict who lived his life on the lam. Jack was a young teenager when he began hopping boxcars across the American West in search of a better life. Along the way, he had the words HARD LUCK tattooed on his fingers, and it was indeed the only luck he would ever know. At the height of the Great Depression, he slept in labor camps and fought for his grub with his fists. While none of his fights ever showed up in the record book, he once claimed to have had 131 of them, "all around the country and in Canada. You just went around from place to place. One day your name was Joe, the next day it was John." He got by on the "two or three bucks" they paid him for the blood he spilled. Marriage to a young woman from Arkansas yielded eight children—four sons and four daughters—yet he and his wife, Arwanda, always remained just a step ahead of insolvency as they roamed from address to address in California. Jack Quarry told journalist Robert Mladinich: "It seems like we were always living at the end of a cotton patch with signs that said, 'Okies and Dogs Keep off the Grass.'" When Jerry was three years old, Jack laced boxing gloves on him. By age sixteen, Jerry had more than a hundred amateur bouts—and that did not include backyard brawls with his brothers. At the barroom, Jack beamed with pride as he spoke of the world championship one of his sons would win one day. Three of the four would turn pro—Jerry, who would become one of the top heavyweights of his era; Mike, who campaigned with some success as a light heavyweight; and Bobby, who fought without distinction as a heavyweight. Matchmaker Don Chargin told me he once saw Jerry tear apart Mike in a sparring session and told Jack, "You have to stop this. Mike is going to get killed." Jack squinted at him and replied: "There's no quit in the Quarrys.'"

White heavyweights had been a prized commodity since the days of Jack Johnson, the audacious black champion who was so reviled by Jack London, the San Francisco Bay Area novelist and journalist, that he organized a bigoted campaign to lure over-the-hill former champion James J. Jeffries out of retirement to challenge him. London told Jeffries, "The White Man must be rescued!" Thus, Jeffries became the first "Great White Hope," of which there would be countless others through the years as black fighters increasingly claimed the championship, beginning with the ascendancy of Joe Louis in the 1930s. While Quarry professed that he was not an advocate of any form of white supremacy, saying he fought only for himself and not any "race or religion," he could just as easily contradict himself, as he had in his comment to the *New York Times* that boxing needed "a white champion to replace Cassius Clay." In an interview with the Associated Press in 1974, he claimed that while it was fine for Ali to "jump around and [call himself] a black militant," it was high time "for the white man to stand up for himself and be proud." Such talk only whetted the appetite of fans who looked to the ring as a place to lay bare racial grievances, yet it was not just the promoters who eagerly exploited it for financial gain. When Quarry outpointed Buster Mathis at Madison Square Garden in May 1969, Frazier congratulated him and said with a sly grin: "Baby, you and me are gonna make a lot of money."

For someone who had earned top grades in school and was said to have found the *New York Times* crossword puzzle to be a snap, Quarry somehow always seemed to be his own worst enemy. Columnist Jim Murray once observed in the *Los Angeles Times* that if Quarry ever penned his autobiography, it would be called *Oops!* At fourteen, his brother Jimmy fractured his arm with a two-by-four. At sixteen, he broke his back when he dived into a swimming pool from a balcony and hit the side; gangrene set in, and he nearly died. At an American Legion baseball game, he shattered a hand when he punched an umpire in the face over a contested call. On another occasion, he collected fourteen stitches when he was clubbed over the head with two pool cues. In the ring, he had shown superior ability as a counterpuncher, yet he seemed to slide into inexplicable periods of lethargy that caused his fans in California to sour on him and question his heart. Although he had thirty-one victories—of which he claimed nine had come on one-punch knockouts—he had been embarrassed by aging Eddie Machen and came up short against Jimmy Ellis in the WBA Tournament finals. Only later did X-rays uncover the fact that he had fought Ellis with three cracked discs, which had occurred during some brotherly horseplay. "The boys are always in a contest of some kind," Jack Quarry told Tom Cushman, of the *Philadelphia Daily News*. "That night Jerry and Jimmy were Indian wrestling, and Jimmy slammed him into a jukebox, back first." Jerry spent eight weeks in a body cast.

Even as Quarry and Frazier professed to be friends at the prefight press briefings, there was an undercurrent of tension between them that dated back three years to their sparring session in Los Angeles at the Main Street Gym, where Frazier had backed Quarry into the ropes and opened a cut below his lower lip. When Frazier had chosen to opt out of the WBA tournament, Jerry called him a "chicken." Eddie Futch flew in from California to work with Frazier during the final ten days of training camp at the Concord Hotel. "Quarry has never faced anyone who fires the bullets Joe does, and as often as Joe does," said Futch, who added that Frazier "hurts you with everything he throws." Upon hearing Quarry boast of the one-punch knockouts he had piled up—and how he planned to dispel any doubts as to the size of his heart when he stepped into the ring with Frazier—Futch observed, "I hope he does think in terms of one punch. If he does, he will get hit with enough leather to cover a sofa." The Garden held a public workout in Times Square for Joe and Jerry a week before the bout, before an audience that included Mayor John V. Lindsay. While not referring to Quarry as a Great White Hope per se, Condon reminded the crowd on hand that he was just that by announcing, "Remember, this is the first Irishman to fight for the *champeenship* since Jimmy Braddock!"

Good sense once again abandoned Quarry on fight night, and Frazier performed precisely the upholstery job that Futch had envisioned. Intent upon proving his bravery, the 12–5 underdog set aside his toolbox and came at Frazier with a hammer. With a Garden crowd of 16,570 squarely behind him, Quarry enveloped Frazier in a vortex of left hooks and chopping right hands in the first round. But Frazier did not back up. The blows caromed off his head, any one of them capable of cutting down a lesser opponent, yet he advanced unimpeded. Near the end of the second round, Frazier caught Quarry with a cruel left hook to his jaw. By the third round, the high energy that Quarry had heaped upon Frazier had dissipated to an occasional spark. Blood began flowing from a cut under his right eye. Through the fourth round, Quarry held his right glove up to protect the wound as Frazier dug punches into his body, each one accompanied by a grunt. "Frazier is a relentless fighter," broadcaster Howard Cosell told his audience from ringside. "He keeps coming. And coming. And coming." As Frazier poured on the pressure in the fifth and sixth rounds, Quarry battled back with the dregs of his resolve, the cut under the eye wider now. Working the head and body, Frazier pinned Quarry to the ropes in the seventh round, then buckled his legs with a left hook to the jaw just before the bell. Dr. Harry Kleiman inspected the eye between rounds and ordered referee Arthur Mercante to stop the fight.

"You wanted blood. Well, you got it," said Jack Quarry, who claimed his son had set out to please the critics who had called him "a bum." My father had been unsparing of Quarry in the pages of *Sports Illustrated* and would again take him to the woodshed in his coverage of the bloodletting at the hands of Frazier, observing that "only a fool has no fear in the ring" and that Quarry had insisted upon being just that in choosing to go to war with Frazier. Co-manager Johnny Flores complained that the fight had been stopped prematurely, despite the fact that Jerry could no longer see out of his swollen eye. "It's a title fight," Flores groused. "He had another eye, didn't he?" Quarry himself had argued with the ring doctor to give him one more round, if only to allow him to hold on to his manhood and go out on his back. He then added, "It was a helluva first round, wasn't it?"

Whatever accounted for the truculence that had come over Frazier in the weeks leading up to the fight, it was gone now. He had earned more than he had in any previous bout—$506,000, once the live gate and closed-circuit revenues were totaled. And he had placed himself in position for an even larger payday against either "Clay"—his preference—or Jimmy Ellis. As Frazier stood in the ring with Cosell for a television interview, exclaiming that he would defend his title against "anybody at any time," Ellis climbed through the ropes in a powder-blue sports jacket.

Cosell spotted him and shouted, "Wait a minute! Here comes Jimmy Ellis! Come in here, Jimmy!"

Frazier turned to walk away and scoffed over his shoulder at Ellis: "You ain't no champ."

Ellis waved at him and replied: "You ain't good enough, boy."

Of the countless places Ali could have chosen to live, he moved from Chicago to Philadelphia in January 1970. With him were his wife Belinda (who in 1975 would be given the Muslim name Khalilah) and their young child, Maryum. Naturally, reporters asked him if he had done so to bug Frazier, if it had been part of some elaborate ploy to hype the emerging rivalry between the two. Ali said no, that he would have rented an apartment in Philadelphia if that had been his aim instead of going to the expense of buying and furnishing a house. He explained that he had come to Philadelphia because it was near New York, where he had frequent business, and eliminated the need to fly. The three-bedroom house, with twenty-two telephones, was located in the Overbook section and was sold to him by one Major Benjamin Coxson, who had befriended Ali two years before and had once been described in the press as "a flamboyant black capitalist with a long criminal record." Even if Ali said he had not intended to annoy Frazier, Joe looked upon his presence as an unwelcome intrusion and warned him to mind his manners. He would abide no "sky larking" from Ali.

Even before the antagonisms between them became entrenched, Joe and Ali were not particularly close. Given the high stakes that were involved and violence that they expected to visit upon one another, it seems fair to say that they could never be, not in the way Ali would profess to be with Coxson. But the two fighters were friendly with each other in a way that rivals sometimes are and understood the immense value they held for each other. George James remembered that Joe and Ali used to drive around the city as they hatched their plans. "They would find a beat-up old wreck and use that so no one would recognize them," said James. Ali talked up the Nation of Islam with Frazier, just as he had done with Sugar Ray Robinson and others. Frazier had no interest, in part because he was a devout Baptist who eschewed racial separatism and in part because of the way the Nation of Islam parted its membership from its money. When Ali was down at heel during his exile, Frazier occasionally slipped him some "love," never forgetting how Floyd Patterson brushed him off when a few hundred dollars would have been the difference between eating and not. Away from the savagery of the ring, Frazier looked upon his peers with a certain esprit de corps and would always lend a hand to one who had come upon hard times, as he would for years in the case of Gypsy Joe. A long conversation between Joe and Ali during a car ride to New York in 1970—when the rancor between them had briefly cooled—was recorded and included verbatim in Ali's autobiography, *The Greatest*. Ali asked Frazier if he would let him come aboard as a sparring partner for two hundred dollars a week in the event that he wasn't allowed back in boxing. Frazier was skeptical.

> **FRAZIER:** First, I like to know, who is gonna be the "sparring partner"?
> **ALI:** Me! I'll be your sparring partner. I'm not fighting. I just said . . .
> **FRAZIER:** Sound like you want to be the main event.
> **ALI:** No. You heard what I was saying.
> **FRAZIER:** I heard you!
> **ALI:** If I get—
> **FRAZIER:** I heard what you said, but to hear you switch it around like *I* would be the sparring partner.

Ali told him he would only do it if they were certain there would be no actual fight between the two. Otherwise, it would "hurt the gate," just as it would not be to their advantage to appear "too friendly." Frazier seemed amused yet wary, never quite sure what cards Ali had up his sleeve. Two years before at the Academy of Music, Ali said, "I look at Joe Frazier and I see ten million dollars." But he had enraged Frazier with his antics that weekend, had gotten him to go for a tire iron, and that played perfectly into the narrative Ali knew he would have to spin in order for him and Frazier to be more than just two men meeting in a boxing ring. Ali understood that white America wanted to see Joe tear his handsome head off. In an evolving tactic, playful in the beginning yet increasingly ugly as their rivalry deepened, Ali would leverage the blind fury he unleashed in Frazier in an effort to ring the cash register. When Ali would call him an Uncle Tom, it was not something he believed—he said so explicitly in his autobiography—yet it was language that incited passions that could be banked and borrowed against. Frazier simply did not have it within himself to play along, not when it turned the black community against him and caused his children to be taunted by their classmates in school. He could not accept being pushed around even in jest, not as a boy in the backwoods of Beaufort and surely not as heavyweight champion.

Even before he purchased the Overbrook house, Ali had been a frequent visitor to the Philadelphia area, where he was never far from the scrutiny of NOI headbanger Jeremiah Shabazz—born Jeremiah Pugh—at Mosque No. 12, at Fifty-Seventh Street and Haverford Avenue. The very same week Frazier claimed a portion of his title by beating Mathis in March 1968, Ali showed up for speaking engagements in Philadelphia and Camden, New Jersey. He addressed an overflow crowd at the University of Pennsylvania, where he said Negroes who favored integration were "heading down the path of destruction." *Philadelphia Inquirer* columnist Joe McGinniss found the talk "so tiresome . . . you find yourself no longer listening, [but] studying instead the organ pipes on the wall." The following evening at the Convention Hall in Camden, Ali again called for the separation of the races and urged the small audience on hand to "cease and desist in the consumption of all pork products," which he claimed were contaminated with "maggots and pus." A confrontation erupted at the door of the Camden event between two Fruit of Islam guards and one Arthur G. Slobodin, twenty-eight, who had been denied admittance. When asked why, Slobodin claimed that one of the guards told him: "Because you are white . . . because you have blue eyes." Police arrested Slobodin and charged him with disorderly conduct "for his own good."

Later that same year Ali became acquainted with Major Coxson, who had had sixteen arrests and twelve convictions in the previous twenty years for crimes including car theft, larceny, weapons offenses, fraud, and interstate transportation of stolen vehicles. But that would not stop him from running for mayor of Camden in 1973. When the *Philadelphia Daily News* asked him if he was concerned that his career as a racketeer would hurt his campaign, the dapper "Maje" replied, "No. The country is run by racketeers." From his purportedly legitimate holdings in an automobile leasing agency and real estate, along with shadowy underworld connections that would ultimately prove his undoing, Coxson had become one of the wealthiest men in the Philadelphia black community. Ali would eventually appoint Coxson as his agent and became his regular houseguest until he moved to Philadelphia. Along with prominent local civil rights leader Stanley Branche, Coxson had big plans for the reclamation of his boon companion and for himself. Whenever the fight between Ali and Frazier came to pass, it would occur in Philadelphia and he would promote it. Ali called him his "gangster."

Initially, the hope was to stage an exhibition bout between Ali and either Frazier or Ellis for charity. Branche applied to the Pennsylvania Athletic Commission for permission on behalf of

Ali, who stated in his formal request that it was his intention "to help the youth of the ghetto remain within the confines of the law . . . talk to various gang leaders," and donate 50 percent to the organizations that formed Black Coalition, Inc., such as the West Philadelphia Branch of the NAACP, the Young Militants, and others. But Cloverlay attorney Bruce Wright told Branche that Frazier had not "a single, blessed thing" to gain by beating Ali in an exhibition and that a loss "would finish his career." Branche accused Wright of being interested in only "the financial end . . . and not the good of the community." In any event, it became a moot point when the commission rejected the request by a 2–1 vote. Upon hearing the news, Ali proposed that he and Frazier have it out on an Indian reservation, which he presumed would not be subject to "government controls." Apparently, the Navajo reservation in Window Rock, Arizona, had approved the bout. Nothing would come of it.

Like some occupying army, Ali camped in close proximity to Frazier during the summer and fall of 1968. When he was not delivering a speech, he was doing a radio or television interview, or popping off in the papers. There he was at the Jet Set Lounge and Bar in Atlantic City. There he was three weeks later in Center City Philadelphia, standing in the back of a yellow convertible rapping with his fans as four attractive young women circulated through the lunch-hour crowd collecting donations for Black Coalition, Inc., laughter escalating as he cried: "Tell Joe Frazier if he's not soon out of town, I'll get Rap Brown to burn him down!" And there he was at the Academy of Music less than a week later, Ali calling out to him, needling him, Frazier furious now, his eye on the tire iron in the trunk. At one point along the line, Frazier glared at Ali and asked, "What you doing in Philly?" And Ali replied, cooing: "To be closer to you, honey." But it was not until the following year that Frazier denounced Ali openly, calling him "yellow . . .

a coward . . . and a disgrace to boxing" prior to the Quarry bout. Ali dismissed Frazier as a "bum" and added, "I ought to give him a good whuppin'."

Was it indeed an act? Was some of it? For onlookers, it was hard to know. Bob Goodman, the public relations man, said no one possessed the promotional acuity of Ali. He slipped in and out of personas as if he were picking costumes out of an old trunk. "But I doubt if Frazier ever was into it," said Goodman. "That was not him." Frazier had a hot button, and when Ali pushed it yet again on a television talk show in Philadelphia on September 22, 1969, the hostility between the two boiled over. Frazier just happened to be tuning in when Ali referred to him as a flat-footed, slow fighter with no class. Frazier called him the following day and challenged him to a showdown at the Twenty-Third PAL (or so Ali stated in his version of events). Ali contacted the local papers and radio stations and told them he would be there by 4 P.M. for what he called "the greatest title fight in the history of a gym." As Ali and Frazier changed into their boxing gear in the locker room, hundreds of people pushed their way into the gym. They stood on windowsills, clung to the banister, and climbed up on the shoulders of others. Hundreds more who could not get in stood outside on Columbia Avenue. Faces pressed against the gym window. Car horns blared as traffic came to a standstill.

"Go out to the park and do your fighting," ordered Police Sergeant Vince Furlong. More police appeared on the scene.

"I came here to rumble!" Ali howled.

Friends surrounded Frazier, holding him back as he exclaimed: "He came here to run me out of my hometown! If I don't take him on now, he'll be trying to run me out of my own house next."

Ali yelled, "He wants to show he can whip me. He says he's the champion. Let him prove it in the ghetto, where the colored folks can see it."

Seventy-five hundred people showed up at Thirty-Third and Diamond Streets, on the edge of Fairmount Park. Or was it ten thousand? As Ali drove up in a red convertible, they were perched in trees, along the fences, and on utility poles. Ali picked up a bullhorn and announced: "This is the biggest thing since the moon walk!" (Which, by the way, had occurred only two months before.) But Frazier was still back at the gym, pissed off at how "Clay" had raised sand in his crib but persuaded by Durham to simmer down. Upon hearing on the radio of the fisticuffs that were brewing, Durham had hopped into his car and sped over to the Twenty-Third PAL. He told Joe to go home, that there would be no brawling with Ali until both of their signatures were on a contract. Durham said, "I would have carried him out on my back before I'd let him fight Clay like that." Word reached Ali at Fairmount Park that Frazier would not be joining him.

"Everybody over here," Ali said, calling the crowd to order. "We just got word that Joe Frazier may not show up. Here he has won all the white man's titles, but he's afraid to fight the real champion."

From somewhere in the rear, a fan shouted, "You tell him, baby."

"So this is your Philadelphia champion," Ali scoffed. "Here I am, haven't had a fight in three years, twenty-five pounds overweight, and Joe Frazier won't show up? What kind of a champ can he be?"

Tensions heightened the following day. Ali and Frazier agreed to do a taping of *The Mike Douglas Show*, during which Douglas sat between them; funnyman Soupy Sales occupied the seat on the end. As Douglas poked and probed with loaded questions intended to stir up some version of the war of words that had occurred the day before, asking Frazier what he called him ("I call him Clay") and did he ever call him Champ ("He was the Champ. I'm the Champ now"), Ali sat with his head down, oddly pensive. When

Frazier said he would fight him any time, Ali replied, "Seven days from the fight, you'll be a weak old ghost." Douglas appeared unsure if he should laugh or dive under his chair. At the end of the segment, Joe and Ali had to be separated from having it out onstage. Was this a gag? Some piece of agreed-upon choreography for daytime television? The studio audience gasped. Outside, the two went at it again before a crowd of bewildered pedestrians. As two members of his entourage held him back, Ali threw a looping right hand that caught Frazier on the shoulder. Frazier stripped out of his jacket and lunged at Ali but was wrapped up by Yank. When Ali threw another right hand, it inadvertently struck Durham, who held his hand up to his eye and yelled, "You crazy mothafucka!" Ali then backed off and someone dragged Joe away.

Joe was livid. That evening, he got in the car with Gypsy Joe and drove out to confront Ali, who was staying with Coxson in the house he would buy four months later. Gypsy Joe gave a version of what happened to his brother, Anthony Molock, and a separate account to my father.

Ali had two Muslims with shoulder arms on either side of him when he came to the door. "My, my, we had some fun today," Ali said. He invited Joe and Gypsy in.

"Right here'll do," Frazier said. "And it weren't no fun for me. Showin' me up like that. Right here in my hometown. Callin' me names."

"Just fun, Joe," Ali replied. "Gotta keep my name out there. Don't mean nothin' by it."

"Coward? Uncle Tom? Only one I've been Tommin' for is you! Names like that ain't just fun. Those sorry-ass Muslims leadin' you on me. It gonna stop right now."

"Don't talk about my religion," Ali said. "I can't let you do that. Go home and cool down."

"Ain't ever gonna be coolin' down now. Fuck your religion.

We're talkin' about me. Who I am." Frazier held out his hand. "This is black. You can't take who I am. You turn on a friend for what? So you impress them Muslim fools, so you be the big man."

Ali said they were done talking and turned back into the house. Frazier called after him, "That's it, get the fuck outta here. Hide behind your shooters. You and me, it's comin'."

CHAPTER SIX

"GIVE ME A HAMBURGER"

Joe and Frank Rizzo, 1971.
Philadelphia Inquirer, staff photo by
Joey Adams

T he White Sands Motel sat at the edge of the Atlantic Ocean in
Margate, New Jersey, hard by the six-story seaside attraction
"Lucy the Elephant." In his room on a Friday evening in the
spring of 1969, prior to his bout with Dave Zyglewicz in Houston,
Joe lounged on the bed in his efficiency apartment and watched
This Is Tom Jones, the weekly variety show starring the Welsh pop
idol who had soared to international fame with chart-toppers such
as "What's New, Pussycat?" and "It's Not Unusual." As women
young and old in the studio audience squealed in something close

to carnal frenzy, Jones leaned over and gave a fan in the front row a peck on the cheek, then turned on his heel and shimmied across the stage. Wearing a black tuxedo instead of his customary splayed collar with a gold pendant nestled in a copse of chest hair, he wiggled his posterior in a way that was so salacious for the day that the camera afforded the viewers at home only an occasional glimpse of his gyrating lower anatomy before it zoomed back up to a less turbulent altitude. Joe scoffed.

"What do women see in him?" Joe asked.

Denise Menz looked up from what she was doing and shrugged. "He's sexy," she said. "Kind of."

"Kind of? How?"

"The way he moves is kind of sexy."

"He calls that dancing?" Joe scoffed. "He should be ashamed of himself. Let me tell ya, he's no Elvis Presley."

Cute and bubbly with a head of red hair, Denise Menz had grown up in Vineland, an agricultural community halfway between the Jersey Shore and Philadelphia. She was the oldest of three children born to John Franklin Menz, a successful bar and restaurant owner, and his wife, Marie, both of whom were by-the-book Catholics. Active in the Cumberland County 4-H program, she won Best in Show as a teenager with her five-year-old quarter horse, Coppertone. "I loved that horse," she would remember. "She would do anything for me. Anything." Upon graduating from Our Lady of Mercy Academy, where she was a prom queen runner-up, Denise enrolled in the Traphagen School of Design in New York City. Off on her own and out from under the scrutiny of her parents, she quickly stopped attending classes, found a job, and fell into "the party life." When she became acquainted with Joe at the City Squire in March 1968, she was immediately taken with him and was quietly thrilled when he invited her to attend the Ramos bout that June at Madison Square Garden. "To show you what I knew then, I thought every fight was held at Madison Square Garden." At the Cloverlay party that was held afterward at the City Squire, she and Joe chatted only briefly. But it was not long before she received an invitation to join him at a Fourth of July party that Durham had planned. It was there that the conversation between the two deepened, and Joe asked her out on a date. She said yes, knowing but not taking into account the implications of his marriage. She would remember that she saw only the twinkle in his eye, saying years later: "I was very young and naïve. You have to remember that."

Casual observers would come to think of her as his girl Friday, doer of assorted secretarial jobs and chief cheerleader. Insiders would know from early on that the relationship had developed into far more than that. But he was careful not to draw attention to it, and she went along. Given how society still frowned upon inter-racial affairs in the 1960s, it would not have gone well for Joe if it became common knowledge that he was engaged in one. Beyond the certain havoc that it would have caused at home—including the potential of a divorce—it surely would have undercut his efforts to build a commercial brand and perhaps impacted his bookings with the Knockouts. "He would have been ruined," Denise said. But reporters during that era seldom strayed into shadowy corners of the lives they covered. When they did, it was only because it was unavoidable, such as when Ali showed up at a press event at the Presidential Palace in the Philippines with his inamorata (and later third wife), Veronica Porche. Second wife Khalilah flew into a jealous rage in his hotel room, just as *New York Times* reporter Dave Anderson showed up for a scheduled appointment with Ali and overheard the commotion from the hallway. When it came to privacy, Frazier was careful not to let his guard down the way Ali often did, which betrayed his creeping paranoia toward the press. Although Frazier had a friendly rapport with the writers and pho-tographers who covered him—genuinely so—he cautioned Denise not to let them get too close.

As the years unfolded, an attachment evolved between them that became progressively more complicated. Early on, she would sit by the phone and anticipate his call each morning when he finished up his roadwork. And it always came. When she was still living with her parents, she would come up with excuses to visit him in Philadelphia. With the help of her father, she secured a job in Center City with a commercial decorator. She moved into a one-bedroom apartment in the small Delaware County community of Colwyn, where Joe would often stop by for dinner. Along with T-bone steaks and fried chicken, she became skilled at preparing classic southern fare, including pig ears and ham hocks. None of it was any good for his blood pressure, which was high and would be an ongoing problem for him through the years. When she overheard him speak of his fondness for chitlins, a soul-food delicacy prepared from pig intestines, she scoured the city in search of a place that sold them. She found them in a grocery store in Southwest Philadelphia. "They came in a gallon box," Denise would remember. "I figured I would cook them up and surprise him."

From that expression of ardor sprang nothing short of a culinary fiasco. According to Denise, the instructions on the box indicated that the contents were "pre-cleaned," so she just assumed that she could prepare them without giving them a thorough scrubbing. As Joe's sister-in-law Miriam Frazier could have told her, given her expertise in the preparation of holiday raccoon, this lapse would prove to be critical. Upon pouring the chitlins into a pot with chopped onions and vinegar, she soon became aware of a horribly foul odor. "I remember thinking, 'How can anyone eat this stuff?'" Denise would say, unaware that the particles that emerged in the simmering broth were not "seasoning," as she supposed, but excrement. As soon as Joe turned into the parking out, he sniffed the air just once and knew exactly what it was. Inside, he opened the pot, grimaced, and asked Denise: "Did you clean these?" Hurriedly, he wrapped up the pot, a pan, and the utensils she had used in a garbage bag, which he dragged outside and heaved into a trash bin. He then sprayed the apartment with a can of deodorizer and lit perfumed candles. On their way out to dinner later, there was a waste management truck in the parking lot, summoned by gagging neighbors who were concerned that there had been a sewer leak. Joe teased her about it for years.

Whenever she would hear that Joe had another bout lined up, she would grow eager with anticipation at the commencement of another training camp. At the Concord Hotel in the Catskills, the White Sands Motel on the Jersey Shore, and elsewhere, she would have him to herself for uninterrupted periods that became increasingly precious. Though Joe usually had a chef in camp, she would occasionally cook for him at the White Sands, where Joe had an efficiency apartment with a kitchen. To pass the hours, they would walk on the beach, throw darts, and spin records from an inventory that included the singers Sam Cooke and Al Green; Dr. King and the Reverend C. L. Franklin; and comedians Jackie "Moms" Mabley and Dewey "Pigmeat" Markham ("Here come da judge!") In keeping with the hoary wisdom that sexual activity robbed a boxer of his legs, Denise claimed that Joe observed a vow of celibacy in the weeks and days leading up to a bout. He had been schooled in this by Durham, who had his fighters double up in hotel rooms on the road and keep the bathroom door open so neither could slip inside and masturbate. "Joe was just so committed to his career," Denise said. "But that was fine with me, as long as I had him with me." He would kneel at his bedside and say his prayers, the way a small child would, and he would read to her from his Bible, interrupting himself during passages to interpret Scripture. "He believed in all of the Ten Commandments except the ones dealing with adultery," said Denise. "He told me, 'The Lord don't care about that.'" When boredom set in, they would occasionally hop in the car and roam the streets in search of another car with which to drag race.

"We were in Margate City on the Jersey Shore during the

off-season, late at night, and we pulled up beside a Plymouth Road Runner at a stoplight," Denise began. "Joe had a Caprice with the biggest engine that Chevy made. He flipped his sunglasses down so no one would recognize him and looked over at the other driver. Joe nodded. The other driver nodded. When the light changed, Joe had no sooner hit the gas than all that was left of the other car was two disappearing taillights. He was gone! Joe looked at me and said, "Damn! Go out tomorrow and buy me a Corvette with four on the floor, a Hurst shifter, and a heavy-duty clutch.' I was laughing so hard I doubled over."

Restlessness invariably got the better of Joe. For as long as those close to him could remember, he had to be on the move, often getting in his car or even a dreaded plane whenever the urge came over him. Although he loved his children and took seriously his obligation to see to their welfare, he felt hemmed in by the sameness of domestic life. Occasionally, as the four walls began closing in on him, he would engage in quarrels at home late at night as a pretext to storm out of the house in search of action. Beyond whatever creative outlet singing provided him, it gave him access to a parallel world free of boundaries, of drinking and laughter and casual assignations with the opposite sex. Women gravitated to him, all manner of women, the way they did to athletes of far lesser stature, and he would oblige them, unheeding of any trapdoors that he might encounter by doing so. Ego coupled with singular intensity had catapulted him from the backwoods of South Carolina to athletic stardom, yet it would be ego alone that impelled him to think that he could conduct himself as he pleased sexually and not pay an extraordinarily high price that would be shared by all. Gloria Hochman, the author and wife of *Philadelphia Daily News* columnist Stan Hochman, would come to know Joe fairly well and wondered if his attraction to Denise had to do in some small part with the fact that it enabled him to "thumb his nose at society and say, in effect:

'If I want to date a white woman, I can do that.'" But Gloria quickly added, "He had genuine affection for her, as she did for him."

Amid the ups and downs of the passing years, Denise would think of herself variously as his confidante, lover, business partner, and, occasionally, indentured servant. Charged by Frazier with running his office, she kept the books, lined up caterers, helped him choose his wardrobe, decorated the gym and the upstairs living quarters, and even did loads of laundry. To the unknowing public, she would have appeared no more than a loyal and dedicated employee. Wherever Joe appeared, it seemed that she was never far away, in later years going behind him and asking some young fighter Joe had just scolded for some slipup: "Can I get you something to eat?" Friends referred to the two of them as George and Gracie, the old comedy duo that featured George Burns as the straight man to the daffy Gracie Allen. "I was George and he was Gracie," said Denise. "We played off each other." Joe appreciated her intelligence and liked her spirit. Unlike other women with whom Joe would have affairs, she chose not to have children with him because he could not offer her a commitment. While he frequently lavished her with gifts, she did not ask for any support other than the small salary that was due her for the job she performed. In fact, in later years she would share with him the money she earned from a seafood carryout business she and her brother Jay owned that later became a three-hundred-seat restaurant and bar. Some who were close to him said that "she was the only one who *gave to* instead of *took from* Joe." Accordingly, Denise said, "I would have done any single thing in the world for him." His daughter Weatta observed, "She loved him unconditionally."

And she would do so always. Even as she understood that she was involved with a married man, she could not bring herself to part from him for any extended period. Her friends would tell her, "Think of all the men you could have." But Denise told them, "I have a Rolls-Royce. Why drive a Volkswagen?" But there would be episodes of turbulence between them with the emergence of other women in his later years, four of whom he had babies with. "It was the only thing we ever fought about," she said. "I knew I was the other woman, but not that there were *other* women." Across the years, there would be periods of separation, but always followed by a reconciliation, with Joe always turning up again and swearing that he would behave. And Denise would forgive him, telling herself that his youth had been foreclosed upon by the early arrival of responsibilities and that he needed his freedom. Only years later would she have him to herself, even if just briefly.

With the 1960s at a close and a new decade at hand, the chaos that had engulfed the heavyweight championship since the expulsion of Muhammad Ali nearly three years before came to an end with the showdown between Joe Frazier and Jimmy Ellis on February 16, 1970, at Madison Square Garden. Given that ring protocol accords the champion preferential treatment, the Garden found itself in something of a quandary in that there were not one but two champions. Financially, they quickly came to terms: both fighters would receive the same guarantee of $150,000 against 30 percent of the total revenues. But who would receive top billing on the show cards—Joe or Jimmy? Who would get first use each day of the training facility at the Felt Forum? Who would get first pick in choosing trunks? Garden boxing director Harry Markson resolved to "split everything down the middle." There would be two posters—one with Joe on top, the other with Jimmy on top. The workout schedule was agreed upon—each moved their preferred time slot by a half an hour to accommodate the other. The dressing-room doors would carry a sign that recognized each as champion.

Jimmy Ellis had lingered for years in the shadow of Ali. They grew up in Louisville on opposite ends of the city, Jimmy on the integrated east side, Ali on the segregated west side. As the two of them came through the amateur ranks together, they became close friends and would remain so, as did their parents and siblings. Along the way, Jimmy and Ali fought each other twice in the amateurs, with each winning one. Ellis won the Golden Gloves and fifty-nine of his sixty-six amateur bouts. Somewhat smaller in build than Ali, he turned pro as a middleweight in 1961, a year after his friend had come back from Rome with the Olympic gold medal. As Ali reeled off a string of easy victories under the Louisville Sponsoring Group and Angelo Dundee, Jimmy struggled under a far less advantageous arrangement with manager and trainer Bud Bruner. By 1964, he had a record of just 15-5, had lost three of his last four bouts, and was on the verge of quitting when he wrote a letter to Dundee in Miami that ended with the plea: H-E-L-P! Angelo agreed to take him on and by a year later had moved him up to the heavies. To throw a spotlight on him, Dundee would remember that he had "the brainstorm" to have him spar with Ali, by then deeply into his conversion to the NOI. Members of the sect leaned on Jimmy to join, according to Jerry Ellis, his younger brother. But Jimmy was the son of a Baptist pastor and sang in the church choir. Jerry told me, "Ali finally told them, 'Hey, man. Leave Jimmy alone.'"

Close to a year and a half had elapsed since Ellis had fought. He had defended his title only once since he had won the WBA Elimination Tournament, beating Floyd Patterson in Stockholm. Hampered by a broken nose in the second round, Ellis eked out a narrow fifteen-round decision over Patterson that Stan Hochman would later claim "had the aroma of limburger." While the Garden angled to set up a title-unification bout between Ellis and Frazier, Dundee seemed in no hurry to take on Joe, if only because there appeared to be far easier paydays on the board to be scooped up. But there would be unforeseen obstacles. Scheduled bouts against Henry Cooper in London, Bob Cleroux in Montreal, and Gregorio

Peralta in Argentina ended up canceled—Cooper due to an injury, Cleroux due to an unexpected loss, and Peralta due to what the promoter called "a total lack of public interest." Dundee moaned, "What do you want me to do? Throw rocks at myself?" As 1969 neared an end, the WBA ordered Ellis to fight Frazier or surrender his title. Contracts were signed that December 29. Frazier eyed Ellis at the press conference in New York and said, "I'm gonna have a ball givin' you the worst lickin' anybody ever got."

Camps commenced in Miami in early January. Frazier had planned to set up headquarters on the Jersey Shore, but a blizzard dropped a foot of snow on the area. As Ellis worked out at the Fifth Street Gym in Miami Beach, Frazier began preparations across Biscayne Bay at the Dinner Key Auditorium, where he had trouble adjusting to the Florida humidity. "I hate it here," he told Denise in a phone conversation. Sparring partners tagged him at will as Yank Durham looked on glumly. "You were pulling away while trying to throw a left hook," he snapped between rounds. "Where'd you learn that?" A report in the *Miami News* claimed that Durham had become "edgy" with how slowly Frazier was shedding the extra pounds he had picked up since the Quarry bout the previous June. He had ballooned to 232 pounds. He got down to 222 by the start of training, but two weeks later had only dropped six pounds, with still another eleven pounds to work off to get down to his 205-pound fighting weight. But no one worked harder or with more enthusiasm to get in shape than Frazier, whose mood improved when he decamped from Florida at the end of January and came north to finish training at the Felt Forum. The cold weather appeared to revive him. One of his sparring partners was Ken Norton, a six-foot-three heavyweight with a 13-0 record who had flown in from California with Eddie Futch. With one eye on the gate and the other on the 4–1 odds that favored Frazier, Garden matchmaker Teddy Brenner contrived to inject a measure of uncertainty in the outcome by spreading the word that Frazier had "overtrained," that Norton and his other sparring partners were hitting him "with everything!" "Look at that!" Brenner exclaimed. "You could stick out your left hand and Frazier would run into it." Told what Brenner had said, Durham roared, "Overtrained? What does that mean?" Futch observed, "He looks about perfect to me at this point."

Eleven days before the bout, Jack Dempsey sauntered into the Felt Forum on a promotional errand on behalf of the Garden and chatted with reporters. Now seventy-four years old, elegantly attired in a blue suit, red sweater, and red-and-gray tie, Dempsey had held the heavyweight championship at the height of the Roaring Twenties, during which he shared space on the American sports pages with a galaxy of stars that included Babe Ruth, Red Grange, and Bill Tilden. As a boy of just fifteen, he saw Jack Johnson train in Reno for his bout against Jim Jeffries in 1910. Nine years later, he battered the behemoth Jess Willard to take the heavyweight championship, prior to which he had some two hundred of what he called "saloon fights" across the Far West. "We had gloves [but] there was no ring," Dempsey told Dave Anderson of the *Times*. "The people would step back and we would go at it until the crowd decided there was a winner. And then they would pass the hat for your money." As Frazier pummeled his sparring partner, Pete [Moleman] Williams, up in the ring, Dempsey looked on with a gleam of appreciation in his eyes.

"He's a rough, tough kid," Dempsey observed. "He can take you out with a punch. He's on top of you."

More than five hundred press credentials were handed out to journalists from across the world. While Dempsey would not venture a prediction on the outcome, former heavyweight James J. Braddock did in a series of articles commissioned by the Associated Press. "The Cinderella Man," who in 1935 had upset even longer odds to win the championship from Max Baer at the old Garden, stood with the underdog Ellis, calling him "the better, smarter, more mature fighter." To offset the pressure that Frazier would apply, Braddock expected Ellis to "move in and out, keep Frazier off balance with a good left hand and always be ready with that 'sneak' right of his." Braddock referred to that right hand as a "deadly punch," and said Ellis "used it better than I have ever seen." But Ali was not sure if Ellis could handle Frazier, even as he picked his old pal Jimmy to win when the press came to him for a prediction. In a conversation with Hochman in early January, during which he seemed to drop his showy dismissal of Joe and revealed his admiration for him, Ali said that Frazier is "too strong." He observed how Ellis had scuffled with Quarry, while Frazier had "annihilated him." Ellis had his hands full with Patterson and had been given the decision, which Ali conceded had been "a robbery." Solemnly, Ali added, "Well, Frazier is much stronger and rougher than Patterson." Durham shared his own prediction with "Two Ton" Tony Galento at the weigh-in: "I'll knock him out in seven." Nodding to the front row, he then added: "I hope you're sitting there. I'll knock him right into your lap."

One by one during the prefight introductions, Galento, Dempsey, and others paraded into the ring and gave a wave to the near-capacity crowd, the old sluggers from a finer day who by their very presence seemed to confer a benediction upon the proceedings at hand. Though Ellis gave away just five pounds to Frazier—at 201, he was three pounds heavier than he had ever fought—he appeared far slighter in build than the sharply contoured Frazier once he shed his robe, his body that of an overfed middleweight. Ellis had said beforehand that he planned to use the ring "like a checkerboard," stepping from side to side in an effort to upset the rhythm of his opponent, who figured to come at him with the throttle wide open. Ellis did just that in the first round of the scheduled fifteen-round bout. Circling the perpetually slow-starting Joe with caution, he scored with left jabs and right leads to the head. Ellis won the round on all three scorecards. From his vantage point in the

corner, Dundee liked what he saw. He would say later, "I thought
I was in like Flynn." Ellis continued to hold his own in the second
round, but Frazier picked up the tempo, driving Jimmy from the
center of the ring into the ropes, where he pummeled Ellis to the
head and body with unanswered left hooks. Frazier sneered, "That
as hard as you can hit, sissy?"

Whatever carefully laid plans Ellis had designed for Frazier
were upended in round 3, as Joe flipped the checkerboard in the air
with a stunning left hook near the end of the round. Ellis held on,
the legs beneath him uncertain as Joe backed him into the ropes.
Announcer Howard Cosell brayed at ringside, "One stupendous
left hook did it!" At the end of the round, Frazier walked back to
his corner with a pleased grin on his face, certain that it was now
only a question of *when*, not *if.* Up off his stool to begin the fourth,
Frazier trapped Ellis in his corner and hammered him with both
hands. Jimmy fell face-first to the canvas. Cosell bellowed: "Down
goes Ellis! Down goes Ellis! He is beaten. Jimmy Ellis is trying to
get up! He is worn to a frazzle! He is a game, game young man!"
Up by the count of eight, Ellis once again found himself engulfed
in a wave of left hooks. Five of them pounded into his body and
head, the last a long, looping wrecking ball that collided with his
jaw. Frazier would later compare it to squarely hitting a baseball
with a bat and sending it "for a ride into the open field." Ellis toppled to the floor. Hysterically, Cosell shrieked: "Oh, a tremendous
left hook! And he cannot be saved by the bell!" As the bell clanged
at the end of the fourth round, referee Tony Perez had come to
the count of six. Three seconds later, Ellis once again labored to
his feet, staggered to his corner and flopped on his stool, where
Dundee worked frantically to revive him.

He soaked him with cold water.

He applied ice to the back of his neck and slipped some down
the front of his trunks.

He waved smelling salts under his nose.

He slapped him on his thighs.

Jimmy just looked back at him with a blank expression. He did not even blink. As Angelo signaled to Perez that he was calling it, Frazier flew into the arms of Durham, who held him suspended from the ground in joy. By then Ellis had come around enough to issue a plea to his corner not to stop it.

"Look, I like Jimmy as a human being," Angelo told the press later. "If I had sent him out there [for the fifth round], he would never be the same. Frazier would have destroyed him. . . . What about his six kids? What about his wife? What about his wonderful mother and father? And what about me? I mean, how do I live with myself?"

Someone at the press conference asked Ellis a question about the second knockdown.

Confused, Ellis said, "There was only one knockdown."

"No. Two, Jimmy," the reporter corrected.

"Oh," Ellis said.

Dundee interjected, "That was why I stopped the fight, gentlemen."

"Free at last!" Frazier boomed as he entered the press conference. Only eleven years had passed since he had come up from Beaufort on the Dog, that bag of fried chicken in his lap. Out of that flabby young man who looked out the bus window with apprehension at the passing scenery had emerged the heavyweight champion, the greatest individual title in sports. With it came a level of acclaim that would have once seemed beyond him, and it was accompanied by a degree of wealth that liberated him from the childhood poverty that had seemed so certain to be his destiny. With the more than four hundred thousand dollars he received for the Ellis bout, he had now earned well in excess of one million dollars. With the help of Denise, he dressed himself at the finest clothiers, given to daring fedoras, long furs, and jewelry galore. Very soon, he planned to move into a sprawling house on the Main

Line, where in the driveway would be parked his Cadillacs, Harley, and other vehicles. Only Ali now stood between him and the only summit yet to scale, but who knew if that would ever happen? Joyfully, Frazier announced that he would quit and become a rock 'n' roll singer unless something with Ali could be worked out. Other than "Clay," Frazier had no one suitable left to fight.

Ali had told the world that he was now retired. To prove it, he had volunteered to hand over the championship belt in the ring to either Frazier or Ellis. But the New York State Athletic Commission rejected his offer. So instead, Ali slipped into a closed-circuit venue in Philadelphia and watched the bout there. When it was over, he stood on a car outside and told the buzzing crowd that Philadelphia was not big enough for two champions. By telephone, he spoke with Frazier later to extend his congratulations, only to get an earful back from Joe. Frazier told him he would stand for no more foolishness, no more shoving and slapping, "the sort of stuff kids do when they are trying to pick a fight." He told Ali to grow up, and perhaps one day they could settle it in the ring. Until then, he told him to stay in his part of Philadelphia, and he would stay in his. Frazier did not reveal what Ali had said in reply, but a week later, when the city held "Joe Frazier Day" in honor of him, there was some speculation that Ali would do something to upend the event. But Ali did not show up. Surrounded by city officials, sports personalities, and assorted guests who included former heavyweight champion Jersey Joe Walcott, Frazier received a proclamation at City Hall, rode though Center City in a twelve-car motorcade, and ended up at a dinner at the Civic Center, where he was given a silver punch bowl with twenty-four cups, one for each of the twenty-four fallen opponents who had paved his way to the championship.

Heads turned whenever Frank Rizzo walked down the street in Philadelphia. In a conservative blue suit, a pressed white shirt, a tie, and shoes polished to a regulation shine, he would leave his desk at the Roundhouse at noon or so and stroll over to the nearby Ben Franklin Hotel, where he would eat lunch each day with a group of underlings on the force. From patrolman to inspector to deputy commissioner to commissioner to mayor, the barrel-chested son of Italian immigrants would polarize the city as he ascended to power, during which he would be known variously as the Big Bambino, the Cisco Kid, the General—and, in less hospitable quarters, Ratso Rizzo. White workingmen and -women looked to him to keep order by cracking heads. Black Philadelphians feared and reviled him, as it was their heads that more often than not ended up cracked. Reporters had a ball sparring with him, once even baiting him into taking a polygraph test. He failed it.

Given his fraught relationship with the black community, Rizzo figured Joe Frazier could be of some help easing any apprehensions toward him as he revved up his mayoral campaign. Rizzo told Joe Hand, "I would like his support. What do you think the chances are?" Hand replied, "Great! He would do that for me." Ever since his sister Mazie steered him to the Twenty-Third PAL, Frazier had developed a fondness for the police and the work they did in the community. Cops had been supportive of him. So he told Hand that he would be happy to join Rizzo for lunch. At the big round table held in reserve for him in the hotel dining room, Rizzo sat surrounded by six of his men, each of whom wore the same blue suit, white shirt, and tie. Frazier looked on as Rizzo scanned the menu and announced, "They've got a great crabmeat sandwich here." He then glanced up at the waiter.

"Let me have the crabmeat sandwich," he said.

With his pad in hand, the waiter then went around the table and jotted down the other orders.

"Crabmeat sandwich," said the first.

"Same," said the second.

"Me, too," said the third.

"Crabmeat sandwich," said the fourth.

"Same here," said the fifth.

"Crabmeat sandwich," said the sixth.

All eyes then fell on Joe. He looked up from his menu and said, "Get me a hamburger."

Hand would remember years later, "The mayor said to me, 'I like that Frazier.' I said, 'I thought you would.' And he said, 'No, all these asses eat what I eat. He had the balls to stand up and say he wanted a hamburger.'"

The old detective held his forefinger and thumb an inch apart and added, "They became that close."

Chances are that Frazier would have had a far different view of his dining companion had his circumstances remained unchanged from his youth. To begin with, he would have been busing the table, not sitting at it. But Frazier would not or could not see beyond the avuncular charm of Rizzo, who presided over a police force that was held in widespread scorn for its brutality. Upon his appointment by Mayor James H. J. Tate as police commissioner in May 1967, Rizzo announced, "Hoodlums have no license to burn and sack Philadelphia in the name of civil liberties and civil rights activities." He labeled his predecessor, Howard Leary, a "gutless bastard" for his handling of the 1964 Columbia Avenue riots; Rizzo had urged him to allow the police to "unholster" their firearms. Although Philadelphia would be spared the havoc that engulfed Detroit (forty dead) and Newark (twenty-five dead) in the summer of 1967, Rizzo found himself at the epicenter of what would be called "a police riot" that November, as thirty-five hundred students protested in front of the Board of Education to demand courses on African American history. At the height of the disturbance, during which twenty-two people were injured and fifty-seven were arrested, the Cisco Kid shouted, "Get their black asses!" Charges of racism would anger Rizzo as the years passed—he promoted blacks on the force and said he counted them among his "friends"—yet in their 1977 book, *The Cop Who Would Be King: The Honorable Frank Rizzo*, authors Joseph R. Daughen and Peter Binzen observed: "Rizzo sought to capitalize on the antagonism that his supporters harbored for black demonstrators. It was a tactic he would use again and again. Ignore the criticism. Attack the critics and their friends. Turn it into an us-against-them situation."

Blacks with arrest warrants became so fearful of injury at the hands of the Philadelphia police that they arranged to turn themselves in to *Philadelphia Daily News* columnist Chuck Stone, a former Tuskegee Airman and speechwriter for the flamboyant Harlem congressman Adam Clayton Powell. From 1972 until 1991, more than seventy-five suspects paraded into the newsroom, checked in at the receptionist desk, and surrendered to Stone, who would summon the police and see to it that no undue harm came to them. Gene Seymour, his nephew, worked as a reporter at the *Daily News* for a period during those years and remembered, "When one of them came in, the newsroom clerk would say, 'Here comes another one.' They would have to find Charlie, who would come in from wherever he had been and accompany the suspect to the Roundhouse for booking." Although Stone said that he and Rizzo had "some riotously fun-filled lunches together" and that he found him to be "charming, generous, gregarious, and a delightfully quick mind," he hammered him as a public servant, calling him "a political exercise in dishonesty, an apologist for police brutality, and a destructive manipulator of city government." While Stone only knew Frazier in passing, he lamented how Joe and Ali had become divided by ideology. Yet he added in 1974: "Joe Frazier knowingly made his choice, and the black community will never let him forget it."

Letters to the *Daily News* during the 1970s came down hard on Frazier as a surrogate for the white establishment. A letter in November 1970 said Frazier "has done nothing to help blacks become shareholders in Cloverlay" and that he could "never be respected by his people like they respect Muhammad Ali." Hand rebutted that in a letter a week later, saying that "many" of the 870 shareholders were black and that Frazier and Cloverlay "do many fine and commendable things in the community to further a deeper love and understanding between the white and black race [*sic*]." When Frazier showed up at campaign events held by Rizzo the following spring, he was excoriated by a reader who claimed, "Joe Frazier is an Uncle Tom campaigning for the white man. His head will be the first Rizzo beats when he gets into office. . . . What is the world coming to?" The reader added that he did not consider Frazier either "a black brother or a soul brother." Raged another reader, "Frazier, you have some nerve to want Rizzo as mayor. Don't you know what he would do to us?" In an interview with Hochman on the subject of Rizzo, Frazier said he had attended some dinners for him "because I consider him a friend. If a black candidate had invited me, I would have gone to his dinners." Frazier argued that he had endorsed no one.

Even if he had not done so officially, it was clear where his loyalties lay. Of the coziness that existed between Frazier and Rizzo, Lester Pelemon observed, "I never liked it, but I would have never told Joe that. He was the boss." Closer to home, his political leanings were met with tongue lashings by his sisters Mazie and Bec, whose daughter Dannette would remember them howling, "You dumb-ass, what are you thinking?" And Joe would reply, "Come on now, none of that." Although Rizzo was then a Democrat (he would later become a Republican), he was unlike any of the Democrats favored through the years by Mazie and Bec, both of whom were savvy and politically active women. In the days when Bec worked as a union organizer in the South, which always seemed to place her in some precarious circumstances, Frazier would occasionally accompany her, if only to keep her clear of harm. He knew the dangers that existed for a black man or woman in the South and had come to understand that it was no better in the North, where he would always sleep with a loaded gun under his pillow. Denise remembered years later, "Joe always had guns around."

Gang activity was on the upswing in Philadelphia in those days, despite the pledge by Rizzo to clean it up. The Black Mafia crime syndicate had established a foothold in the city, specializing in drugs and extortion, and would operate in conjunction with the Nation of Islam out of Mosque No. 12 on North Broad Street, only a few short blocks from where Cloverlay opened up the gym for Joe in the spring of 1970. Due to what his business manager Gene Kilroy called a tendency to be "gullible and naïve," Ali brushed up against these criminal elements during the three years he lived in Philadelphia and Cherry Hill, New Jersey. Major Coxson—whom Ali had jokingly referred to as his "gangster"—would be gunned down with his wife and two children by members of the Black Mafia during a home invasion in Cherry Hill over an apparent drug deal gone wrong, less than a mile from where Ali had purchased a house from Coxson. Headed in Philadelphia by Jeremiah Shabazz, the NOI had been at odds with Coxson over his chummy relationship with Ali, despite the fact that Ali remained under a one-year suspension by Elijah Muhammad for his March 1969 announcement on *The Tonight Show* that he was willing to fight for money again.

Even as Mosque No. 12 would become a haven for young boxers who lined their pockets with the spoils of criminal endeavor—the trainer George James said he worked with a handful of them— Frazier was spared any interference from Shabazz or his ilk, perhaps to some extent because of the "firewall" that Joe had created by aligning himself with Rizzo. According to Frank Rizzo Jr., "Franny," who later served for sixteen years as a Republican member of the Philadelphia City Council, his father would call Joe if he or one of his men spotted him with the wrong people and say, "Hey, Champ. You gotta act like the Champ." Rizzo hated Shabazz, and years later he would accuse him of being behind a fire on South Street that had been set during a robbery by elements of the Black Mafia. Joe Hand said Rizzo confronted Shabazz in a restaurant and told him, "I know what you did, and I am going to get you for it, you son of a bitch." Hand would remember that Rizzo had a Black Panther agitator hauled into the Roundhouse, sat him down in a chair, and barked, "If I ever again see you cross that bridge from New Jersey, those boots you are wearing are going to be filled with blood." Hand said simply, "Nothing was going to happen to Joe." Of Joe's relationship with Rizzo, Pastor Kenneth Doe added: "His momma Dolly would probably have said, 'The Lord put that white man there to protect my boy.'"

Beyond whatever protection he received from Rizzo—which included the loan of his personal bodyguards as needed—Frazier seemed to enjoy the easy access he had to power. Franny told me it was not uncommon for Joe to drop in unannounced to see his father. He remembered, "Even if Dad was in a meeting, he would stop and say to whoever was in his office, 'Would you like to say hello to Joe Frazier?' And he would wave Joe in." Franny believes that Frazier certainly helped his father politically in the black community. "He had a lot of influence among the row-house folks, preachers, corporate executives," said Franny. "Joe would tell them, 'You don't know him the way I do.'" While Frazier would always say that he did not favor one political party over the other, he had a bias toward law-and-order candidates and, over the years, lent his support to Republican presidential candidates Richard Nixon, Ronald Reagan, and George H. W. Bush. Contrary to the prevailing belief, he found himself sharing the same political space as Ali, who migrated across the years from a confirmed racial separatist in the 1960s to a malleable advocate for conservative causes. In fact, Ali even appeared to be on the same page as Rizzo, telling Hochman in 1975: "I like what I hear about him. I like his looks." He did not hold Rizzo accountable for not "cleaning up the black neighborhoods," adding: "People got to clean up their own neighborhoods."

By virtue of his friendship with Rizzo, Frazier came to know Nixon during the first term of his presidency. Hand said that when Nixon came to Philadelphia, he used to huddle up alone with Rizzo inside his helicopter. Nixon was so enamored with how Rizzo had cracked down on crime that he eyed him to head the Federal Bureau of Investigation when J. Edgar Hoover died in 1972. According to Denise, Joe had encountered Nixon at an event in Washington in 1969 or thereabouts and the subject of "Clay" came up. Nixon asked him if it would be helpful to him if he arranged it that "Clay" could fight again. Joe said that it would. Joe told Denise that Nixon then asked, "Do you think you can beat that motherfucker?" Joe looked at him in disbelief and replied, "Yes, sir."

———

Of the ten members on the Cloverlay board of directors, only Arthur Kaufmann stood opposed to a showdown between Joe and "Clay." From his position as a top executive of Gimbels for twenty-three years, Kaufmann became a leading figure in the city, serving as president of the Chamber of Commerce and cochairman of the local Olympic Committee. He had become acquainted with Frazier during his gold-medal quest and was the largest shareholder in Cloverlay. Although Kaufmann had not served in uniform himself, he worked as the civilian aide to the secretary of the army from 1954 to 1979 and had deep connections within the Pentagon, once appearing on the same dais with General William Westmoreland at the Valley Forge Military Academy. Consequently, he looked upon "Clay" with antipathy, and could not sit still as Joe "tarnished" his wholesome, All-American image by consorting with a draft dodger. Anchored behind the desk of his flag-draped office in Center City, Kaufmann told *Philadelphia Evening Bulletin* columnist Sandy Grady: "If Clay wants to fight, let him fight the Viet Cong."

Contractually, Kaufmann had no say in who Frazier fought. Yank Durham decided that, and his authority was inviolable. But

Kaufmann remained firm in his belief that Frazier should not fight "Clay" until he served his hitch in the army or a jail sentence. When the Cloverlay board held a vote in November 1970 to approve a bout against "Clay" that had been proposed for Florida, which appeared to have the backing of Governor Claude R. Kirk, Kaufmann cast the only dissenting ballot. He told Grady: "I cannot condone anyone, whether it be the world heavyweight champion or Joe Doakes, flagrantly violating the law of the land, and especially a law having to do with the security of the country. . . . The dollars Joe Frazier receives will never restore his good reputation or his own self-respect." Kaufmann stepped down from the board. Neither Joe nor Durham weighed in publicly with comment on Kaufmann, yet Cloverlay attorney Bruce Wright did say that if it came to it, Joe would fight "Clay" in whatever prison to which they sent him. Hand remembered, "Even having a vote was silly. Joe fought who Yank said he would fight. Period."

Early efforts to find an agreeable host city for Ali-Frazier were blocked. When it was proposed in August 1968 to bring it to Albuquerque, New Mexico, boxing commissioner Tim Kelleher said he would only approve it if he received the okay from General Lewis Hershey, the director of the Selective Service. Kelleher told the *Albuquerque Journal:* "You get Hershey to write us a letter saying this is a good idea and I'll buy the first ticket." Up in Philadelphia, Coxson held a press conference at the Bellevue Stratford a year later saying that he had procured a license for his friend Ali from the Mississippi Athletic Association and that the bout would be held on December 15 at the Jackson Coliseum. Upon hearing that, Frazier replied, "Is he crazy? What am I gonna do down there? How am I even going to get there? Even the Greyhounds don't go down to Mississippi." Florida was in the running with an intriguing proposal by Murray Woroner, who had come to some small fame by producing a computer-simulated bout that year between Ali and Rocky Marciano. Woroner envisioned Ali and Frazier facing each other in an actual bout in a film studio in South Miami, with only the press and perhaps two thousand spectators in attendance. The bout would be broadcast to closed-circuit venues worldwide.

Eagerly, Wright observed, "This could be the shape of the future." Although he had been working with Dundee to set up the unification match with Ellis, he quickly set that aside in anticipation of the "dream bout" with Clay in February or March 1970. Upon hearing of the negotiations that were taking place, Governor Kirk said he loved the idea of bringing the fight to Florida. He added that he would love it even more if it were held in Tampa instead of South Miami, which encouraged Wright to withdraw from stalled talks with Woroner. Although Kirk had no legal jurisdiction over who received boxing licenses in Florida—those decisions were up to the commissions in each city—the young Republican who in the 1960 presidential campaign had headed up Floridians for Nixon observed: "Florida is the sports capital of the world, and this fight will only help to enhance its image." But the Tampa City Council came out against it, as did four Florida congressmen and the *Tampa Tribune*, which carried an editorial opposing it under the headline: WE OBJECT—CONSCIENTIOUSLY. Nor were the citizens of Orlando too pleased when it was proposed to bring the bout there. Thus, forty-eight hours after giving the fight his blessing, Kirk announced that he "heard from enough people in Florida to know that they do not want to have the fight take place here." Two months later, Frazier stepped into the ring with Ellis and unified the championship.

With Ali unavailable and no other immediate challengers lined up, Frazier headed off to Las Vegas, where he and the Knockouts were booked into a lounge at Caesars Palace for a two-week engagement. Sharing the bill with him were singer Nancy Wilson and comedians Godfrey Cambridge and Norm Crosby. At a 10 P.M. show, near the end of his run, Frazier was in the second chorus of "Knock on Wood" when his feet flew out from under him as he executed a split. "They must've waxed the floor or something," Frazier said. With a throbbing right ankle, he came back out for the 2 A.M. show before going to the emergency room, where it was learned that he had a fractured ankle. Chagrined that he was unable to squeeze into his "groovy suit," he was back onstage forty-eight hours later in a full leg cast. He sat on a stool as he sang. Crosby quipped: "I've worked with a lot of great casts, but this is ridiculous." Doctors told him he would have to wear the cast for six weeks.

"The itching drove him crazy," Denise said. "I had to stick a coat hanger in there and scratch it. And he would say, 'No, no, no. Over more!' He had it only two and a half weeks when he told me to cut it off. He said, 'The doctors overreacted.' To loosen it up, he stuck his leg in the shower. I got in there with a serrated bread knife and began sawing it. The wetter it got, the more the plaster came apart. Finally, with him pulling at it and me cutting it, it came off. I rubbed his leg down with a loofah for a half hour and applied Jergens lotion to it. He got up and walked like nothing had ever happened."

Joe joined the Knockouts for some concert dates in South Carolina and then followed Sammy Davis Jr. into the Latin Casino in Cherry Hill. With the assistance of Patti LaBelle and the Bluebelles, who opened for him and pitched in as backup singers, Frazier performed two shows nightly before crowds that were so sparse that *Daily News* columnist Hochman observed that "a guy could get snow-blind staring at the empty tablecloths. . . . Which is too bad, because Frazier the singer is like Frazier the fighter, busy all the time, holding nothing back. And how many honest, lay-it-on-the-line guys are around today?" Toward the end of his two-week engagement, Frazier could not conceal his disappointment at the poor turnouts. He told *Philadelphia Inquirer* writer Jack Lloyd that he knew people looked upon him as "another athlete going into show business, just a gimmick to bring people out and disappoint them." But he remained hopeful that he could prove to people that he was more than that. To accomplish that, he said he and the Knockouts should go back into the studio and record some of their own stuff. He planned to create his own label.

"Who knows, I might not ever put on the gloves again," he told Lloyd. "I mean that. But then the chance to fight Clay might come along and I would say, 'Yeah.'"

But when and where would that happen? Ali himself was of the belief that he would remain sidelined until the White House intervened on his behalf. Even if he was unaware of how Nixon had referred to him in his conversation with Frazier, he knew it was unlikely that he would receive any help from Washington. Toronto emerged as a contender for the bout, yet Ali had been stripped of his passport and would be unable to gain approval of his petition to leave the United States for eighteen hours. Some interest was expressed in taking the fight to the state of Washington, but the Washington State Athletic Commission voted it down 2–1. "Forty-nine other states say no; why should we stick our necks out?" asked W. B. "Red" Reese, one of the dissenting commissioners. Michigan was in the running and had the apparent support of boxing commissioner Chuck Davey, a former welterweight contender. But Governor William G. Milliken withheld his approval. As these and other sites were debated, *Philadelphia Evening Bulletin* writer George Kiseda laid into Philadelphia for not taking initiative to claim the event, particularly since both Frazier and Ali were taxpayers. Kiseda excoriated Governor Milton Shapp, Mayor Tate, the chamber of commerce, and the Convention and Tourist Bureau.

"These people spend a lot of words and money pretending that Philadelphia is one of the great resorts of the world when actually Joke City is the last resort anybody would consider for a vacation, a weekend or even a night out," Kiseda wrote. "Philadelphia entertainment? You can go to Linton's and count the scrambled egg sandwiches that come down the conveyor belt."

Atlanta stepped to the fore in the quest to reel in Ali-Frazier by granting Ali a boxing license in August, with the hope of staging the bout in the five-thousand-seat City Auditorium on October 26. Georgia state senator Leroy Johnson had arranged the license under the nose of arch-segregationist governor Lester Maddox and with no apparent behind-the-scenes aid from Nixon. But Durham did not consider Atlanta to be a suitable site. Nor did he think Philadelphia would do, even if Kiseda had inspired civic effort to get the event with his sharp rebuke. As Frazier prepared to face hard-hitting light heavyweight champion Bob Foster later in the fall, Ali used the Georgia license to begin working himself back in shape by taking on number-one contender Jerry Quarry, whom he stopped in three rounds with a cut over his left eye that required eleven sutures. Back in Philadelphia, Major Coxson held a party for Ali a few days later at his supper club, Rolls-Royce, where Ali sat in a corner booth in an Edwardian suit as fans approached him for autographs. Councilman George X. Schwartz looked on and observed, "This is Frazier's town. He won't like this." To which City Council President Paul D'Ortona replied, "No more. This is a Muhammad Ali town from now on."

Only 5,914 fans showed up to see Frazier face Bob Foster on November 18 at Cobo Arena in Detroit, which had originally reserved the date and venue for an Ali-Frazier bout. Frazier gave away four inches in height and five and a half inches in reach to the light heavyweight champion, yet outweighed him on paper by twenty-one pounds, despite the fact that his trainer Bill Gore said Foster had consumed twelve cans of beer each day for a month in order add extra pounds. According to publicist Bob Goodman, Foster actually weighed less than his announced weight of 188 pounds. "Going into the weigh-in, I placed a ten-pound weight under the belt of his trunks," said Goodman. "His actual weight was 178." Notwithstanding the considerable weight disparity, Foster had called it "just another fight" and pricked Joe by calling him "a dumb fighter" in the daily press, noting "he fights the same way every fight." Futch later said the comment irked Frazier, who to the disbelief of Foster came out in the first round intent upon proving the scope of his boxing skills. The long-limbed Foster caught his stocky opponent with a jab to the chin and followed it up with a right hand that Frazier later said dazed him. "Man, he was rattling my brains," Joe told *Sports Illustrated.* When Frazier came back to his corner at the end of the round, Durham roared, "What did I tell you? Didn't I tell you, you couldn't give this tall, skinny guy punching room, he'll knock your fucking brains out?" Goodman would say that Durham told Joe, "You gotta jump in his pants. You gotta bite his chest."

From that point on, all Foster would remember was that Frazier bore down on him like a big train. Early in the second round, Frazier floored him with a sudden left hook to the head. Foster would not remember beating the count, nor the double left hook to the head and body with which Joe finished the job. Goodman would remember, "Joe hit him so hard that Bob sprained his ankle as the punch spun him around." As referee Tom Briscoe counted over him, Foster struggled to rise, getting up to one knee before toppling back to the canvas. Durham cradled his head in his lap and cursed Briscoe for not stopping it sooner.

Foster remained in a soupy fog back in his dressing room, where he began slipping back on the boxing shoes that one of his trainers had just removed.

"What are you doing?" the trainer asked.

"I'm getting ready to fight," Foster replied.

"Bobby, the fight's over with."

"What do you mean?"

"He knocked you out."

Foster blinked back tears and asked, "On national TV? Oh God, everybody saw me get knocked out."

Ali continued to gleefully disparage Frazier as "a tramp, not

the champ." In the aftermath of the Foster bout, Ali had described
Frazier as "just a machine. He has no skills. He has no sense." Yap,
yap, yap; did Ali ever stop to take a breath? Frazier would say that
Ali talked whenever he was "afraid" before a fight, but "this time
he is gonna be real scared." When Ali held his final tune-up on
December 7, 1970, at Madison Square Garden against Oscar
Bonavena (New York had relented and given him a license on the
heels of his comeback in Atlanta), Frazier watched on closed circuit
from Monticello Raceway, thinking, as Ali found himself battered
across fifteen rounds by the ungainly Oscar: *Don't blow it, man.*
Ali won the fight by technical knockout when he floored Bonavena
three times in the fifteenth round. While it was precisely the type
of challenging effort that Ali and Dundee were looking for, Frazier
was far from impressed with what he had seen. "The way I saw
it, Bonavena won every round," Frazier told reporters. "All those
mistakes Clay made tonight, he'll make again. And when he does,
look out."

Christmas was a joyous event that year. Joe, Florence, and their
five children moved out of their small house on Ogontz Avenue
and into a deluxe seven-bedroom house on two and a half acres
in Montgomery County, which would enable him to avoid paying the onerous city wage tax on his purse from the upcoming Ali
fight; Ali would do the same by moving from Philadelphia into a
house across the river in Cherry Hill. To celebrate the occasion
of Frazier's move, some sixty-five relatives came up from Beaufort
and elsewhere for a reunion. Gone were the days when the bare
walls during the holidays were decorated with pictures torn from
catalogs, of luxurious possessions others had but the Fraziers could
only dream of. Now, the accoutrements of wealth were arrayed before the family's very eyes in a house that had seven bedrooms, four
bathrooms, three recreation rooms, and a six-car garage; Joe would
later have an outdoor pool installed in the shape of a boxing glove.
As Dolly Frazier sat darning socks by a color television set, the

house was alive with the laughter of her assorted grandchildren, all of whom had come to look upon Joe as someone they could count on for guidance and encouragement. One of them was his nephew Mark, who would remember driving in a car with him once. Ali had come to his school assembly in New York and had yanked him out of the audience for some playful sparring when he learned that he and Frazier were related.

"He said he's gonna whup you," Mark said in the car. "All the kids think so, too."

Joe glanced over at the boy seated in the passenger seat and replied, "Son, he don't want any part of Uncle Billy."

CHAPTER SEVEN

THE FIGHT OF THE CENTURY

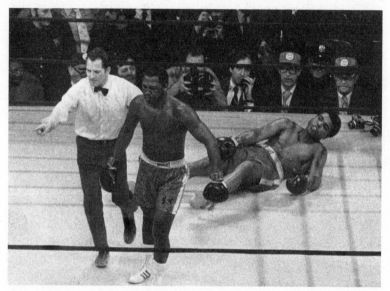

The Fight of the Century. March 8, 1971. AP Images

C loaked in the privacy afforded by a tranquil Sunday morning, Harry Markson and Teddy Brenner stepped off their train at the elevated station in North Philadelphia and strolled across Broad Street to the Cloverlay Gym for an appointment with Joe and Yank Durham. From his first fight in the old Madison Square Garden, against Dick Wipperman in March 1966, Frazier had enjoyed a relationship with Markson and Brenner that had enriched all. The Garden had become his professional home and had done well by him, effectively creating a state championship that enabled him to bypass the WBA elimination tournament and unify the title that had been stripped from Ali by beating its winner, Jimmy Ellis. With Ali now back and licensed to fight in New York, Markson and Brenner had come to Philadelphia with an offer to secure the promotional and site rights for what would go down in history as the Fight of the Century.

A pad of paper sat on the table in front of each as they convened in an upstairs office at the gym that December day in 1970. On them would appear a number that would have once seemed beyond belief to Joe, one with such a long trail of zeros that he was uncertain how to write it down; he would always say math was never his "thing." According to Brenner in his book *Only the Ring Was Square*, he and Markson tendered a guarantee of $1.25 million against 35 percent of the gross; they were prepared to offer Ali the same deal. Given their projections of the revenue streams, Brenner explained that both Frazier and Ali could end up earning far more than the guarantee, perhaps as much as "$6 million and $7 million." But Durham had been approached with a competing offer, one that was less speculative and had a guarantee attached to it that was even more eye-popping: $2.5 million.

"Who made the offer?" Markson asked.

Durham replied, "Jerry Perenchio."

Suitors for the event had been stepping forward with gusto, only to find themselves either outbid or outmaneuvered. Early on, it appeared that it would end up at the Astrodome in Houston, where Top Rank, Inc. founder Fred Hoffeinz offered the same arrangement as the Garden but with a capacity of thirty thousand more seats. Former New York Jets owner Sonny Werblin also appeared to have the inside position at one point, in partnership with NBC and Johnny Carson. But the forty-year-old Beverly Hills talent agent Perenchio prevailed. In London on December 15, he had heard that the event could be had for five million dollars, so he ran up a sixteen-thousand-dollar phone bill at his hotel in calling seventy possible backers. One of his lawyers recommended that he pool his efforts with Jack Kent Cooke, the Los Angeles sportsman who himself had been exploring the possibility of bringing the event to his arena in Inglewood, California, the "Fabulous Forum"; he also owned the NBA Lakers and NHL Kings, and had held a 25 percent stake in the NFL Washington Redskins. The two talked and were, as Cooke would later say, "compatible spirits." Although the event would be held not at the Forum but at Madison Square Garden, which had the support of both Durham and Chauncey Eskridge, the attorney for Ali, Cooke agreed to authorize a letter of credit for $4.5 million, with the additional five hundred thousand dollars paid as a site fee by the Garden. Cooke and Perenchio agreed on a 60-40 split in favor of Cooke, while the Garden received 30 percent of the live gate and closed circuit in New York and Illinois. The $2.5 million flat-fee guarantee for each fighter was a staggering sum for the day—$15,217,531 in 2018 dollars—larger than any athlete at that point had ever been paid for a single performance.

By his own admission, Perenchio knew nothing of boxing. But as the head of the talent agency Chartwell Artists, he had cultivated an appreciation for star power, numbering among his clients Marlon Brando, Glen Campbell, Richard Burton, Elizabeth Taylor, and Andy Williams. In hyping a promotion that he claimed would "transcend boxing," Perenchio said that while both Joe and Ali were "superstars," the "focus is on [Ali]. Without him in it, it would be like seeing *Gone with the Wind* without Clark Gable." Of course, Ali was in full accord with that, telling the *Wall Street Journal*: "Why do you think the fight is so big? Frazier never wrote any poems. He never did any shuffles." Perenchio estimated worldwide gross revenues of between $20 million and $30 million, which included $1.2 million from the live gate, $15 million from closed circuit, $4 million from between-round commercials, $1.5 million from foreign radio and television, $1 million from program sales, and $10 million from a documentary film that Perenchio planned to release. To add to the take, he also planned to auction off the gloves, trunks, and shoes of both fighters, noting: "If they can sell Judy Garland's *Wizard of Oz* red slippers for fifteen thousand dollars, then we should get at least as much for these." The avaricious young entrepreneur added that he expected the trunks to go for even more if either had blood splattered on them.

A press conference to announce the fight was held on December 30 at Toots Shor's, the Manhattan saloon that for years had been a hideaway for Hollywood celebrities and ballplayers. Although it had been scheduled for noon, it was an hour late getting started due to an incident that had occurred back at Madison Square Garden as the contracts were being signed. In an act of playful exuberance, Ali clutched Joe and began tussling with him. As Joe Hand stepped in to break it up, Joe tore the seam in the back of his brown plaid suit, which necessitated a return to the City Squire Motor Inn and a change into a blue suit. Ordinarily, given the care Frazier took with his appearance, it was the type of behavior that would have incensed him. But Ali could have given him a hotfoot on this day and Joe would have done a tap dance. As he told his brother Bozo, "How can I be mad at a guy who's making me this much money?" Once the press conference finally got under way, Ali sat at the table, separated from Frazier by Edwin Dooley of the New York Athletic Commission, and shouted for all to hear across the five boroughs and beyond: "Believe in me, not the newspaper writers. Joe Frazier is a fraud, an amateur. I'm the real champion! Frazier is a flat-footed machine. I'm an artist." On and on he harangued, until Frazier got up from his seat, handed the artist a wedge of lemon, and told him, "Suck on this for a while."

Eight weeks before the March 8 bout, Joe opened up training camp at the Concord Hotel in the Catskills, where the conditions were perfectly fine for Alpine skiing but somewhat less so for getting into shape for a prizefight. Far from the sunshine and palm trees Ali enjoyed in Miami Beach, where he toiled each day at the Fifth Street Gym and clowned with the visiting press, Frazier braved subzero temperatures as he set out each morning before sunup to do his roadwork, his body wrapped in layers of clothing as a hard wind whipped down from the hillsides. Even with steel-studded boots, he found himself slipping on patches of ice. By the end of his run, icicles had become tangled in his beard. "I had to scrape them off with a comb," Denise would remember. Seventeen inches of fresh snow in early February drove him back to Philadelphia, where he set up headquarters at the Franklin Motor Inn and did his running each morning in Fairmount Park in more agreeable wind chills. Occasionally as he ran, he would picture "Clay" at a post on the path ahead of him. "I gotta run and get him," he said. "So I run hard to that post and then maybe he'll be at the next post and I gotta run after him again."

Controversy greeted Frazier as he arrived back in Philadelphia. The Southern Christian Leadership Council had called a press conference in January and threatened to call for a boycott of the bout unless certain conditions were agreed upon. Chiefly, the SCLC asked that closed-circuit ticket prices be lowered from twenty dollars to ten dollars to accommodate the stressed incomes in the black community; that black businessmen be allowed to control 20 percent of the closed-circuit locations; and that 15 percent of the revenues in each state be used in the ongoing effort to battle drugs and crime. "Tell 'em to drop dead," Durham told Hochman in the *Daily News*. In searching out "the best possible deal" he could for Frazier, Durham said he opened up the bidding to all parties; if the SCLC had a desire to be involved, "all they had to do was come up with the money." Now that it had been sold to Perenchio and Cooke, Durham recommended that the group take it up with them, not Frazier. Some weeks later, an offshoot of the SCLC calling itself the Consumer Education Protective Agency set up a picket line on the sidewalk in front of the gym with placards that proclaimed: GET RICH QUICK FIGHT PROMOTERS EXPLOITING BLACK COMMUNITY! LOWER ALI-FRAZIER TICKET PRICES. Upon crossing the line with one of his security guards, Frazier just shook his head and said, "Sometimes silence is the best answer." But Durham was far from silent with one of the organizers.

"Where were you people when I was working on the railroad breaking my back as a welder?" Durham boomed. "Where were you people when Joe Frazier was scraping out a hundred dollars a week working in a slaughterhouse?"

Stan Hochman later asked if Durham had any sympathy for the needs of impoverished black people. The *Daily News* columnist observed that Durham clenched his fists for emphasis and replied, "I'll tell you something. If I give money to fight poverty, it goes to fight poverty among whites as well as poverty among blacks."

Standing-room-only crowds packed the Cloverlay Gym to watch Frazier train, as his eleven-year-old son Marvis sat at the door and collected a dollar per head. Loud soul and R&B music rattled the windows as Frazier worked over his four sparring partners, Ray Anderson, Ken Norton, Don Warner, and Moleman Williams. At one point along the way, Frazier sidelined the 175-pound Anderson by opening a cut on his upper lip with a crushing left hook. The wound required three stitches to close. Although none of the sparring partners possessed the same skill set as Ali, Durham claimed that they represented four riddles for Joe to solve each day. J Russell Peltz, then a twenty-four-year-old promoter who was just getting his start, remembered that he dropped by the gym one day, paid his dollar, and asked if he could use Warner on a boxing card he had coming up at the Blue Horizon, once the fraternal lodge of the Loyal Order of Moose but now a boxing arena that aficionados of the sport would fondly characterize as a "bucket of blood." Durham eyed Peltz and scoffed: "I pay my sparring partners more than you can pay them for a fight." Peltz laughed years later as he told me, "And he does this in front of hundreds of people."

Calling Peltz off to the side for a chat, Durham told him he had fifteen complimentary tickets or so for the fight that he wanted him to sell on the street. He explained to Peltz that whatever he could get for them above face value, they would split 50-50. Had he been more sophisticated, Peltz later said, he would have understood the small fortune he could have reaped. But he was still young and inexperienced and he let the tickets go one by one at face value until he got down to a final front-row seat for $150. He considered keeping it for himself but would remember, "I was not exactly setting the world on fire then." He sold the ringside seat to the owner of the Tippin' Bar at Broad and South Streets—"a guy we called Booster."

Peltz circled back to the gym and handed five hundred dollars or so in cash to Durham, who apparently had been expecting a far healthier roll. "He looked a little disappointed," Peltz said. But the young promoter still planned to attend the bout. When the Garden announced that it was opening up sales for twenty-dollar seats, Peltz and two friends took a train from Philadelphia to New York the evening before to stand in line. As he would remember it, it was cold and snowy; they took turns going for coffee. Finally, the Garden box office opened up at 8 A.M. "You were only allowed to buy two tickets each, so I took two," Peltz said. One of them he would use for himself; the other he would leave at the front desk of the Americana Hotel for a woman who did not show up to use it. Years later, he wished that he had retrieved it and held on to it, noting: "Do you know what an unused Ali-Frazier ticket goes for today?"

Yank Durham always expected trouble to walk in the door. To appreciate his steep level of paranoia, he once spotted the bewhiskered artist LeRoy Neiman in the gym with a sketch pad. He called Joe Hand Sr. over and asked irritably, "What is that guy doing in here?" When Hand explained who he was, and that Cloverlay planned to commission him to do artwork to display on the walls of the gym, Durham replied sharply, "Then why he is over there taking notes?" In the way royalty once feared poisoning by a conniving member of the court, Durham never allowed Joe to take a drink of water or a bite of food in a restaurant unless it was first sampled by someone on his security staff. To be on the safe side, Durham would cook for Joe himself unless he knew the chef personally. With an eye always peeled for some disruption of the norm, he found himself faced with exactly that in the weeks leading up to the showdown with Ali: there was a spy in camp.

Someone had been calling Ali each night at ten thirty to give him a rundown on how Joe was faring in his workouts. Ali himself told reporters he had "a friend" feeding him information. "His sparring partners are beating up on him," Ali said. "The other day, one of them bloodied his nose and hit him with over a hundred right hands." As inside information goes, none of it held any compelling value, yet Ali seemed to think it gave him some small psychological edge, saying: "If I can get the edge on Frazier, why not do it? Football and basketball coaches send scouts to watch their opponents." For his part, Frazier appeared bored by the whole subject. But Durham pursued it as if he were a bloodhound prowling through the underbrush on the heels of an escaped con. For two weeks, he had his eye on Don Warner as his leading suspect. Warner had been beaten by Cassius Clay in February 1962 on a fourth-round technical knockout and was known to be connected to the Nation of Islam. Although he denied being a member, Warner admitted to occasionally attending meetings. (He later adopted the name Hasan Muhammad.) Durham summoned him to his room at the Franklin Motor Inn.

"You rotten cocksucker," he roared. "I talked to Clay last night and he said you are the spy." He told Warner to go to the Cloverlay office and pick up his final check. He was through. But when Warner went to the bank and tried to cash the check, someone had stopped payment on it. He guessed it was Durham. Warner denied any wrongdoing, saying, "Joe did more for me than Clay ever could dream of doing. . . . I have a lot of respect for Joe."

Guilty or not of the espionage that Durham alleged, Warner found himself engulfed in the choosing up of sides that surrounded the fight, which played itself out in combative letters to the editorial pages of the papers and even in actual shootouts between partisans in bars. As word spread in the city that Durham had found his man, Warner became a target of hostility. Crank calls began flooding the telephone line at his house, where he lived with his pregnant wife and their children; sometimes the voice on the other end would cuss him out, sometimes he would hear just a click and a dial tone. One evening when he was sitting in his living room, there was a sudden splintering of glass. Someone had thrown a rock through his window. Attached to it was a note that said: "spy." As he moved through the city on his daily errands, he could feel the accusatory eyes of strangers upon him. Warner later filed a ten-million-dollar lawsuit against Frazier, Durham, and Cloverlay, claiming that his "good name, fame and reputation has been brought into disgrace and disrepute" by his dismissal. Frazier ended up settling it for a relatively small sum.

Someone told Ali that his spy had been caught and asked, "What are you going to do about reports now?"

Ali said that he still had two other "friends" feeding him information. And winked.

To sprinkle the event with some showbiz glitter, Perenchio called upon an old Hollywood connection, actor Burt Lancaster, to work as host of the closed-circuit broadcast with Don Dunphy, the veteran blow-by-blow announcer, and Archie Moore, the former light heavyweight champion. For a publicity stunt, it was announced that Ali and Frazier were each to spar a round or so with Lancaster, whose only pugilistic experience had been that he played a boxer in the 1946 film *The Killers*. Supposedly, Frank Sinatra was also to be on hand and work the corner for Lancaster. But the announced event never came off. The crowd that packed the Cloverlay Gym on February 16 was disappointed to learn that Ali and Lancaster were in Miami Beach; no one knew where Sinatra was. When Lancaster did show up in Philadelphia later that week, it was not to don the gloves but to chat with writers; he said he abandoned the plan to spar, not because of any fear for his safety but because he considered it "a cheap gag." Wearing a wide-brimmed black cowboy hat— a prop from his forthcoming film *Lawman* that he later gave to Frazier, which prompted Durham to call Joe "Dark Gable"—Lancaster accompanied Joe and Yank to the card that J Russell Peltz held at the Blue Horizon on February 22. Only blocks from his gym on North Broad Street, Frazier walked into the arena and found himself swallowed up in boos as the crowd began chanting: "ALI, ALI, ALI!!"

Even though Frazier said, "What do I care if people holler for him?" whatever forbearance he had projected toward Ali on the day the contracts were signed had eroded, as Ali portrayed Frazier as a favorite of the white establishment that had persecuted him. With a voice that carried over the oceans, he labeled Frazier an Uncle Tom in the way others had done to Louis Armstrong and Sammy Davis Jr. In an article that appeared in *Life*, Frazier excoriated "Clay" as a "clown" and "phony." According to correspondent Thomas Thompson, Frazier spoke "with a cold bitterness, a rancor that was surprising for him." Joe claimed that "Clay" had been "brainwashed . . . and now he has all the black people brainwashed, too." Frazier snarled, "Clay called me an Uncle Tom. You going to tell me Clay don't have white friends? What color is his trainer, for example?" In other interviews with the press, he said that "Clay" had separated him from his people, and that the two of them "could never be friends." When a reporter told him that Ali planned to seal his prediction in an envelope and asked him, "Would you want to take a peek?" Joe frowned and snapped, "What do I want with his mail? I get three hundred letters a day myself. Besides, I doubt if he can write that well anyway."

Beyond a brief period during which he seemed to lack zip, in part later explained by the fact that he was skipping breakfast, Frazier sailed through his training with grim efficiency. Usually, he would be in bed by 8 P.M. and awake by 2 A.M., some three hours before he typically did his roadwork. Either he would wake up his crew and run anyway or he would lie in bed and think of March 8 and "Clay." Given the three-and-a-half-year layoff Ali had recently come off, Frazier was certain that he no longer had the same bounce in his legs that had once allowed him to glide out of trouble. With the help of assistant trainer Eddie Futch, who had a granular eye for the imperfections in Ali and would later beat him with Ken Norton, Frazier blocked out a strategy that would enable him to use his superior conditioning to wear down Ali over several rounds. As he had with his previous opponents, he would come at him with far greater pressure than Ali had ever faced. To have any chance of winning, he would have to cut off the ring on Ali, encourage him to take chances by showing him an exaggerated bob-and-weave, slip inside that long jab that Ali discharged in $^4/_{100}$ of a second, and then batter him with hooks along his beltline. The objective would be to force Ali to "punch up, not down," which is to say, close the space with which he had to work and get him into position to use only his uppercut and hook instead of that extraordinary jab.

Tensions climbed with the emergence of death threats. Frazier received a letter that warned him to "lose or else," which was followed by a phone call a few days later to the same effect. Who was behind them? Had they had some connection to the Nation of Islam and its shadowy underground? Or had it been just some fan with a grievance known only to himself? Although it appeared that Frazier was unfazed by any of the threats that came in—in fact, he laughed them off—Les Pelemon and Durham were concerned. "We thought of it as kind of serious," said Pelemon. Frank Rizzo assigned a security detail of four Philadelphia detectives to join Joe for the drive to New York. When they emerged from the Lincoln

Tunnel, they were joined by a squad of four NYPD detectives that included Joseph Coffey, who in later years would gain fame by his work on the Son of Sam case and his arrest of Mafia don John Gotti. Instead of escorting the caravan to the City Squire Motor Inn on Seventh Avenue between Fifty-First and Fifty-Second Streets—which had received a bomb threat that would prove unfounded—Coffey and his men accompanied them to the Pierre on Sixty-First Street. On the eve of the bout, according to his autobiography, Frazier was watching a crime show, *The Naked City*, when his telephone rang.

"Joe Frazier, you ready?" Ali asked from the Hotel New Yorker.

"I'm ready, brother," Frazier replied.

"I'm ready, too, Joe Frazier. And you can't beat me. 'Cause I am the greatest."

"You know what? You preach that you're one of God's men. Well, we'll see whose corner the Lord will be in."

"You sure you're not scared, Joe Frazier?"

"Scared of what I'm going to do to you."

"Ain't nothing you *can* do. 'Cause I'll be pecking and poking and pouring water on your smokin'. Bye, Joe Frazier. See you tomorrow night."

"I'll be there. Don't be late."

———

Early on March 8 in his suite at the Pierre, Joe turned to Lester Pelemon and noted, "Well, Puff, it's the countdown. At this time tomorrow, it'll be over and we can go on the road singing again." Up at 8:30 A.M., Frazier passed up breakfast in anticipation of the weigh-in and shadowboxed until he worked up a sweat. Casually, he spoke of heading down to South Carolina, of kicking back with the Knockouts there and buying a new house for his mother. Surrounded by a phalanx of eight detectives, he showed up for the weigh-in just before noon, wearing a vivid green and gold brocade robe. Stitched on the back of it were the names of five of his children. He removed it and stepped on the scale: 205½ pounds; Ali would weigh in later at 215 pounds. For the security of the two men, the Garden provided them both private rooms where they were fed and shielded from the commotion that only intensified with the onset of evening, as New York society and others who aspired to it stepped from the rear doors of black limousines parked six and eight deep from the curb. As Pelemon would remember, Frazier no sooner checked into the room that had been set aside for him than he was "ready to go." He began pacing the floor.

"Keep him calm, Les," Durham told Pelemon. To pass the hours, Pelemon and Joe began singing some of the songs from their act a cappella—"Proud Mary," "Knock on Wood," and "My Way." Four hours before he was due in the ring, he reported to his dressing room. There, he shadowboxed some and then lay back on a padded table, chatting with Durham, Futch, and Pelemon. Slowly, he stripped out of his street clothes and began donning his gear: first his athletic supporter, then his trunks and shoes. Seated on a chair, he held out his hands for Durham to wrap with layers of gauze and adhesive tape as a representative from the opposing dressing room looked over his shoulder. As Durham did the same while Ali had his hands bandaged, Ali looked up at him and asked if he thought Frazier could whip him. Durham would remember, "I told him I KNEW Frazier was gonna whip him."

Durham returned to Joe's dressing room, walked him to a corner, braced his hands on his shoulders, and looked deep into his eyes. "Well, we're here," he began. "I want you to know what you've done, boy. There will never be another Joe Frazier. They all laughed. You got us here. There's not another human being who ever lived I'd want to send out there, not even Joe Louis. Win tonight, and the road will be paved in gold. Think of those mammy-suckin' white people and the hot fields soaking up the sweat and hope of your parents. You were made for this moment. Take it, cocksucker!"

"Five minutes!" someone shouted.

Frazier lowered himself to one knee and prayed aloud:

"God, let me survive this night.

"God, protect my family.

"God, grant me strength.

"And God . . . allow me to kick the shit out of this mothafucker."

From his twenty-dollar seat among the hoi polloi in the upper tier of the Garden, J Russell Peltz placed his binoculars up to his eyes and scanned ringside: sure enough, there was Booster, the Philadelphia bar owner to whom he'd sold a ticket, standing cheek by jowl with Frank Sinatra, there with his new Leica as a photographer on assignment with *Life* (which used one of his action shots the following week on the cover). Had he turned his glasses elsewhere, he would have spied a galaxy of stars that included Elizabeth Taylor and Richard Burton, Gene Kelly, Ethel Kennedy, Lorne Greene, Joe Namath, Marcello Mastroianni, Hubert Humphrey, Michael Caine, Abbie Hoffman, Bernadette Devlin, Hugh Hefner, and scores of others. In the hours prior to the bout, the ringside aisles had become a runway of eclectic fashion, with tuxedos and formal evening dress tossed together with furs, hot pants, and stiletto heels in a display of what Joe Flaherty of *The Village Voice* called "the horny sheet sharers of sex and ego." Of the 760 press credentials the Garden issued (they denied another five hundred), the lineup along press row included Norman Mailer (*Life*), Budd Schulberg (*Playboy*), and William Saroyan (*True*). In setting the scene in the piece he filed that night for *Sports Illustrated*, my father observed of Ali in his lead: "He has always wanted the world as his audience, wanted the kind of attention that few men in history ever receive. So on Monday night it was his, all of it, the intense hate and love of his own nation, the singular concentration and concern of multitudes in every corner of the earth, all of it suddenly blowing across a squared patch of light like a relentless wind."

Roaring cheers from the crowd of 20,455 accompanied Ali and then Frazier as they came into the arena and ducked through the ropes. Across the globe, thirty million viewers tuned in on the closed-circuit hookup, the largest television audience in history. Richard Nixon had it piped into the White House; Ali had joked in the weeks prior that the president would probably call Frazier to congratulate him if he won, but never Ali. Covered in a long red robe with white trim, Ali bobbed up and down on his toes, sliding this way and that. At one point, he brushed by Frazier and said, "Chump." Frazier scowled at him. When referee Arthur Mercante summoned the two men and their seconds to ring center for the prefight instructions, they stood with their eyes locked on one another in a hard glare, Ali yapping, Frazier nodding and grinning. According to *Newsweek*, Ali warned Frazier, "Look out, nigger. I'm gonna kill ya."

That was precisely what Ali set out to do as the bell summoned them to action. Aware in some part of himself that the three and a half years away had taken something from him, even if it was unclear exactly what, Ali intended to capitalize on the vulnerability Frazier had displayed during the early rounds of his previous bouts. As Frazier advanced toward him in the first round, Ali stood flat-footed and hammered him with the first seven blows, including two solid left hooks. With a three-and-a-half-inch height advantage and six-and-a-half-inch edge in reach, Ali peppered Frazier with jabs in the second round, which opened up room for him to punch down on Frazier. But just as he had vowed that he would, Frazier closed the space between the two, taking blow upon blow from Ali in order to get inside and work his body. Near the end of the round, Frazier backed Ali into the ropes and tagged him with a clean left hook to the head. Ali turned to the crowd and shook his head, "No," as if to say that the blow had not had any effect. From ringside, announcer Don Dunphy observed: "Frazier has been hit with solid right hands that would have felled an ordinary man."

But Frazier was no ordinary man on this evening. Down low in that bob-and-weave, he drove Ali to the ropes in the third round and caught him with a solid left to the head. Ali used his superior speed to beat Frazier to the punch in their exchanges in close, but Frazier landed a textbook left hook that staggered Ali. He covered up on the ropes and again shook his head "no" to the crowd; he would do that often. As Joe pressed the action again in the fourth, Ali punished him with a volley of combinations, then walloped him with a triple hook to the head. "Joe is taking a battering," exclaimed Dunphy. "Anyone else would be on the floor." But Joe battled back and unleashed a violent body attack, closing the round with a solid left hook to the head just before the bell. As Angelo Dundee leaned over him with energetic instruction between rounds, Ali sat on his stool and gulped for air.

Some reconstituted version of the old Ali was back in the fifth round. Up on his toes and moving side to side, he created space between himself and Frazier and sent out that jab. But he was soon down off the balls of his feet, as Frazier backed him to the ropes and whacked him with body shots. As Frazier fired away, Ali appeared to be tiring somewhat, his arms now perceptibly heavier. He caught Joe with a stiff one-two to the head, whereupon Frazier dropped his hands, pointed his left glove at Ali, and laughed, as the former champion himself had taunted opponents in his younger days. "I wanted him to know that I could take everything he had and then some," Frazier said later. Ali backpedaled as Frazier followed, his hands still down as Ali swung at him wildly. At the bell, Joe shook his glove in the air in defiance. Ali saw it out of the corner of his eye and flinched. Although in his commentary Dunphy had observed that Ali was "piling up points," Frazier was ahead on all three scorecards.

Only hours before, Ali had opened his sealed envelope and revealed his prediction that he would take out Frazier in the sixth round. Ali later conceded to fight film collector Jim Jacobs that he had only prepared for a six-round bout; Ali also said, "I hit him with everything in the fifth round and he didn't fall. . . . I knew I was in for a long night." With Ali against the ropes in the sixth round, Frazier nailed him with not one but two left hooks to the side of the head. Ali spun free but Joe again backed him into the ropes, where he pounded Ali along the beltline again and again. Dunphy observed, "Ali is almost like a sitting duck." Content to lean back on the top strand of the ropes, Ali peppered Frazier with taps to the top of the head as he allowed Frazier to whale away at him, certain that Frazier would punch himself out and fold up in exhaustion. Jacobs said that Ali fought 60 percent of the bout with his back to the ropes.

Frazier was still on top of Ali in round 7. When they were tangled in a clinch, Ali asked Joe, "Don't you know I'm God?" Frazier replied, "Well, God, you're gonna get whipped tonight." As his fans joined in chanting "ALI! ALI! ALI!" Ali once again began throwing heavy leather at Joe in the eighth. Frazier again forced him to the ropes. Ali stood there, his hands at his sides, as Joe teed off on him, the crowd now chanting, "JOE! JOE! JOE!" When Ali returned to the corner at the end of the round, Dundee laid into him. "Stop playing!" he said. "Do you want to blow this fight? Do you want to blow everything? You're giving away rounds and letting him build not only a lead but his confidence."

Moving again in round 9, Ali clipped Frazier in the head with his jab, as Joe appeared now to be tiring himself. But Frazier once again backed Ali into the ropes and strafed him with a left hook to the head, then another to the body. When Ali extricated himself, Frazier pursued him to ring center and landed another left hook to the head. The blow wobbled Ali, whose eyes appeared to lose focus. Ali battled back by throwing wild rights as Frazier continued to advance on him. But just when it appeared that Ali was tottering on the edge of a cliff, he somehow found a second wind. Ali let go a quartet of left hooks, three of which landed solidly and staggered

Joe. Ali winged him with combinations and then a left hook to the head as the round came to a close.

Early in round 10, Joe drove Ali to the ropes again and scored heavily to the body, only to be accidentally poked in the eye by Mercante as he was separating them from a clinch. Joe turned away briefly, looked over his shoulder, and gesticulated angrily as Durham cursed Mercante from the corner. Seizing upon this opportunity, Ali began firing off more combinations. From his corner, Dundee shouted, "Go, baby, go!" Just before the end of the round, Ali poked his head out of a clinch and shouted out to the crowd, "He's out!" As the two men found their corners at the close of the tenth, Frazier was now ahead on only two of the three scorecards.

Joe took command in round 11. With Ali backed up against the ropes, Joe lunged at him with a left hook to the head that dazed Ali. Quickly, Joe followed with a left hook to the chest and another to the head. Ali began to sag to the canvas. Certain that Ali would go down for the count, Joe turned toward a neutral corner, only to turn back again and discover that Ali had abruptly straightened himself. Slack-jawed yet now clear-eyed, Ali backpedaled in an exaggerated wobble, his hands down at his sides, as if inviting Joe to come in and finish the job. Was Ali actually hurt? Or was he luring Frazier into a trap? Across the ring, Frazier held his own hands down at his sides and balked, unsure if Ali was laying a con on him. Futch would tease Joe for years for not pressing his advantage at that point and taking what he called "The Long Walk" across the canvas to reengage Ali.

Ali sat down on his stool in his corner at the end of the eleventh, where Dundee pleaded with him over the din in their ears. "Get off the ropes! Use your jab! Get back on your toes!" But Ali did only some of that in the twelfth and not much more in the thirteenth, as his jaw began to show signs of swelling. Frazier pinned Ali against the ropes and pounded on him. Ali staged a rally in the fourteenth, during which he pushed Frazier back with an array of hooks and combinations. Physically, Frazier looked as if he had been in a car crash and had gone through the windshield. His face was an abstraction of bruises and welts, his left eye now beginning to close.

Pushed by each other to the very edge of their endurance, Ali and Frazier lugged what remained of their strength and stamina to ring center and touched gloves as the fifteenth and final round commenced. Ali went to work quickly, throwing combinations to the head, as Frazier backed him to the ropes. Twenty seconds into the round, Frazier doubled up with a left hook to the body and the head. The first blow struck Ali on the elbow; the second one caught him on the jaw, jacked him into the air as his legs flew out from under him, the red tassels on his shoes fluttering under the hot ring lights. The crowd roared.

Quickly, Ali got up at the count of two. Dundee would later say, "He was out when he was hit and he was up when he hit the floor." The roar grew louder. When Mercante concluded giving Ali the mandatory eight-count, Frazier rushed in to finish the job, firing off shots to the head and body. Frazier tagged Ali with another left hook to the jaw that caused Dunphy to gasp, "Ohhhh, what a shot." But Ali would not go down again. With Joe on top of him, Ali finished the round, even staging a small rally as it drew to a close. At the bell, the roar grew louder still.

Overwhelmed with pride at how far they had come, Durham cradled Frazier in his big arms as a sea of commentators, cameramen, and boxing officials filled the ring in anticipation of the decision; Joe had the appearance of a frail child. Someone slipped a robe on Ali as Dundee snipped the laces of his gloves. Then announcer Johnny Addie stepped to the center of the ring and said, "Ladies and gentleman . . . Referee Arthur Mercante scores it eight to six, one round even, for Frazier. Artie Aidala scores it nine to six for Frazier." The crowd erupted in wild cheers, as Ali lowered his head. "Bill Recht has it eleven rounds for Frazier, four rounds for Ali. The winner by unanimous decision and STILL heavyweight champion of the world, Joe Frazier!"

Joe stopped to see Ali in his dressing room. Seizing upon the physical ruin he had visited upon Frazier, Ali murmured, "You're beautiful. You're the real champ for now." Ali was then taken to Flower Fifth Avenue Hospital to have his jaw X-rayed; he would not talk to the press until the following morning, when he sat in a curved-backed, upholstered parlor chair at the Hotel New Yorker and claimed he won "nine rounds." He pointed out that the referee and two judges who voted against him had been appointed by the very same commission that had stripped him of his title. Of his fifteenth-round knockdown, he observed: "I saw the hook coming and I figured I would ride with it. But it was hard." He added that he had no recollection of falling. "Boom, it was that quick," he said. In terms of the overall importance of the defeat he had just suffered, he minimized it, explaining to Robert Lipsyte, of the *New York Times:* "News don't last too long. Planes crash, ninety people die, it's not news more than a day after. . . . World go on."

The previous night, after the fight, cameras had whirred and clicked as Joe sat at his press conference and answered questions, his soggy green robe draped over his shoulder as beads of sweat popped out on his forehead. Welts had risen over both his eyes. Dry blood appeared alongside his left nostril. Pelemon held an ice bag to his swollen left cheek; Durham fed him pieces of chipped ice. Of Ali, he said, "He takes some punch. That shot I hit him with, I went down home and got that one. From out in the country." Joe called him "a good man" yet thought Ali should apologize for the ghetto slurs with which he had showered him. Joe said that when Ali told him during prefight instructions, "Look out, nigger, I'm gonna kill ya," he replied, "That's what you're gonna have to do." Although Frazier was skeptical that Ali would even want a rematch, he claimed he would be all for it if the occasion arose. First, though, he planned to take some time off, perhaps even a year. "Me and Yank, we got to go home and take it easy for a while. Live a little," he said. And then added: "God knows I whipped him."

———

Only a few days later, Gene Kilroy, Ali's business manager, was with Ali at the Essex House in New York when he received a call from the writer Budd Schulberg. "Gene," Schulberg said, "I just heard that Joe Frazier has died." Stunned, Kilroy looked over at Ali and passed along what Schulberg had told him. "What?" Ali replied, his eyes wide with consternation. When Kilroy hung up with Schulberg, Ali told him, "I will never fight again." To confirm the report, Kilroy called Frazier's physician, Dr. James C. Giuffre, at St. Luke's and Children's Hospital in Philadelphia. Giuffre told him that Frazier had been admitted to the hospital with high blood pressure and related issues. The doctor said "it had been a little touch and go" at first, but that Joe was doing better and would be fine. Kilroy relayed the good news to Ali, who replied. "All praises to Allah. I'm glad he's okay." Kilroy said later, "Ali really cared for Joe."

Beyond his horrific facial wounds, Frazier was experiencing internal issues that only worsened in the hours and days that followed. Amid the chaos in the ring prior to the announcement of the decision, Don Dunphy reported, "Joe Frazier seems to be sick in his corner." According to Durham, he and Frazier went back to the dressing room and finished off a bottle of champagne between them. At 1:45 A.M., they dropped by the victory party at the Statler Hilton, which drew one thousand revelers at thirty-five dollars a head and where Florence sat at a table signing autographs. As Duke Ellington and His Orchestra played in the ballroom, Joe found a secluded spot in the hotel kitchen, where he sat behind a pair of sunglasses and sipped from a glass of water; he confessed to a United Press International reporter that he was feeling "weak." Later that Tuesday, after Florence had gone back to Philadelphia, he opened a bottle of champagne with Denise at the City Squire. "We did it, baby!" he said. And with a laugh, he took two long pulls and immediately threw it back up. Denise had never before seen him have trouble holding down alcohol. "He could eat nails and wash it down with white lightning and be fine," Denise would say. "So yes, I could see he was not himself."

For as long as Denise knew him, Joe did not call attention to any sicknesses that occasionally befell him, nor did he have any tolerance for even tender concern by others over the physical damage boxing caused. Whenever Florence or Denise would express any worry to that end, he would snap: "Cut the bullshit. This is what I do for a living; I don't want to hear any more about it." But in the days following the Ali bout, he would exhibit an array of alarming symptoms: elevated blood pressure, acute nausea, severe headaches, and a lowered pulse rate. On that Tuesday evening, Denise was so upset that she called Durham, who in turn summoned Dr. Edwin Campbell, the physician for the New York State Athletic Commission. Campbell would remember later that Frazier was "more exhausted than any fighter I had ever seen." Upon giving him a cursory exam in his hotel room, he recommended that he order something light from room service and perhaps consider an enema. "A what?" Joe replied, his voice rising. Denise remembered, "Of course, he wanted no part of that. But I did order him some scrambled eggs and bacon." By Wednesday evening, he could not keep any food down, urinate, or even walk. Again, she called Durham, who once again sent for Campbell. According to Denise, the doctor gave her "two red pills" and said, "Here, give these to him." Not keen on taking drugs of any kind, Joe flushed them down the toilet. On Thursday evening, she placed a call to the Cloverlay suite in the hotel in search of Joe's brother Tommy. She hoped he could spark Durham into action and get Joe to the hospital. Durham was still celebrating hard when Tommy told him, "Either you take him to the hospital or I am." Denise would later say, "Oh yeah, they were all still drinking. Even the bellhops were loaded."

Quickly, Durham hopped on the phone. For privacy, Campbell recommended that they go to a hospital in the Catskills, far away from any nosy reporters who would splash his condition across the front page. But Durham got ahold of Dr. Giuffre and arranged to sneak him into St. Luke's. By car, they sped down the New Jersey Turnpike, dropped Denise off at her apartment, and proceeded to the hospital. Handing Joe off to Giuffre, Durham headed to catch a plane to the United Kingdom, where he was scheduled to do the color commentary for the Joe Bugner–Henry Cooper bout that Tuesday. Only Joe Hand Sr. stayed with Frazier during the harrowing weekend that followed. "His blood pressure was so high, they were concerned he was going to have a stroke," said Hand, who sat by his bedside. "Doctors were running in and out of his room. His blood pressure would go up and then it would go down." Hand prayed: *Let him live.* In his autobiography, Frazier would remember that Giuffre sent him to the recovery room, where he was placed on a bed of ice. "I stayed like that all night. Was it a dream that a spirit took my hand in the wee hours and told me I'd be okay? To this day I can recall His presence. His touch on my hand." According to his daily diary, Richard Nixon placed a call from the White House that Saturday at 2:42 P.M. but did not get through.

Cloverlay denied that Frazier was in the hospital when first contacted by the papers. When asked why Joe had more or less dropped out of public view—he had canceled an interview with Howard Cosell and some other appearances—a spokesman said he had merely come down with a case of the flu. But when a reporter spotted Frazier on Monday in an anteroom at the hospital eating lunch, Giuffre provided a more accurate evaluation of his condition. He said that Frazier had a blood-pressure reading of 180/90, which he seemed to think had more to do with "tension" than any issue that emerged from the blows he had endured. Stable enough to leave the hospital on Monday, Joe stopped in to see Rizzo at his campaign headquarters, where the General wrapped him in a hug, then spent the evening at his home in Montgomery County. There, he took some aspirin for a headache and later took some more when he was unable to sleep. When he showed up at the hospital the following day, ostensibly to show films of some of his old fights to the patients in the drug clinic, Frazier appeared "sluggish and washed out," according to Dr. Giuffre. To get his blood pressure under control—it was once again abnormally high—the doctor admitted him for a battery of tests and placed him on an intravenous feeding tube. Along with his elevated blood pressure, he had "transient hematuria" (blood in the urine). He would remain in the hospital for twelve days.

Churning amid the scuttlebutt that circulated during this period—which included the speculation by a British doctor that Ali appeared to have been doped during the bout, perhaps without knowing it (Ali dismissed this as "silly"), it was rumored that Frazier had suffered a detached retina in his left eye at the hands of Ali. While that rumor would prove to be unfounded, he had been having trouble with his vision since he came back from the Olympics in 1964. According to Dr. Myron Yanoff, a Philadelphia-based ophthalmologist who examined him in July 1971, he had congenital cataracts. The condition was in its early stages, not yet affecting his right eye but "slightly more advanced with a mild decrease of his vision in his left eye." Yanoff said he apprised Frazier of "the potential worsening of his cataracts with his continued boxing" and "the risks to his vision." Even the small deficiency he had in his left eye at that point explained the undefended shots he took from Ramos, Ellis, and Ali, who threw long right hands that were delivered at a speed and an angle that were hard for Joe to pick up. As his vision in both eyes declined in the ensuing years, he would have corrective surgery, buy contact lenses by the boxful, and, as Gypsy Joe before him had done, pass his eye tests by memorizing the charts. None of this would be publicly known during his career.

As Frazier recovered in the hospital, United Press International broke a story that cast grave concern over his prognosis. The source of it was Dr. Campbell, who speculated with a reporter that based on his examination of Joe in New York, he could very well have suffered a severe head injury. According to Campbell, the symptoms Frazier presented were consistent with either a concussion or a subdural hematoma (a blood clot between the outer two layers of the brain). Upon hearing this off-the-cuff diagnosis, Dr. Giuffre scoffed and dismissed it entirely, saying that Joe had none of the symptoms that suggested either. Although he had been experiencing headaches, they were neither persistent nor accompanied by vertigo or a dilation of the pupils. Within a day of the report, Campbell claimed he had been misquoted and added that there had been "no findings present to indicate that [Frazier] suffered a severe head injury." Had there been, Campbell said, he never would have allowed him to leave New York without having him evaluated in a hospital. Campbell said Frazier was "suffering from severe emotional and physical exhaustion."

Visitors streamed in and out of his suite. Doctors stopped by to have their pictures taken with him in his faux-tiger-skin robe. Fruit and flower arrangements came in, which Joe dispersed to the nurses on the floor. Friends dropped by, often with *their* friends. Telephone calls flooded the hospital switchboard, including one from Bob Hope. For a special scheduled to air in early April, Hope asked if he could tape a segment with Joe in Philadelphia; Frazier told him he would be happy to do it. Hundreds of cards, letters, and telegrams poured in from across the world. A Cloverlay secretary stopped in and said they had received a call from someone at Columbia Pictures, who proposed doing a biopic that he claimed would be certain to gross $20 million ($115 million in 2018 dollars). To battle boredom, Joe sat across the two hospital beds that had been pushed together for him and strummed his guitar. As his blood pressure came down and he began to feel better, he would slip out of the hospital to buy clothes, do a television interview, or attend a car show; he even swung up to New York to watch the Knockouts perform and drove back the same evening. An emergency exit door in the stairwell at the hospital was always left ajar for Denise, who would show up in the evening with a piece of American Tourister luggage packed with fried chicken, her special potato salad, and the apple turnovers Joe liked. She would leave before the day shift came in at 7 A.M.

Upon discharging Joe from the hospital on March 27, Giuffre pronounced him "fully recovered." In the near term, Frazier planned to embark on a trip down to Beaufort to find his mother a new house and was discussing a possible trip to Europe with the Knockouts in May or June. Although Durham had raised the possibility that Frazier would retire—"What does he still have to prove? Why push it?"—Joe was not yet prepared to walk away from boxing, particularly with the big money that was still out there to be scooped up. But he had no immediate plans to step back into the ring, if only because of his tax liability. Once his taxes, expenses, and payments to shareholders were accounted for, Frazier would clear something like $750,000 from his $2.5 million. Both he and Ali were aghast at the tax bite they incurred, including $350,000 each from the State of New York. By virtue of their 60-40 split, Cooke cleared $450,000, while Perenchio took home $350,000. While Cooke would entertain grandiose visions of a rematch at the Forum in L.A., Frazier told reporters that a second fight would happen only on his terms. Of his asking price of five million dollars, he told reporters that there would be no 50-50 split with Ali. It would be, as he put it, "All mine."

Had he fought in a less complicated era, chances are he would have had that leverage. But what Frazier would find and could not bring himself to accept—not then, not ever—was that it was not enough to just be champion. Even in defeat—perhaps especially so—Ali had galvanized his base, which looked upon him as something more than just an athlete. In his column in the *Philadelphia Evening Bulletin* the day after the fight, Sandy Grady observed: "Ali lives too deeply at the heart of our troubles to vanish." Literal tears were shed at his undoing by Frazier, whom writer Hunter S. Thompson chopped into pieces in his book *Fear and Loathing in Las Vegas*. For Thompson, the outcome was "a very painful experience in every way." According to Thompson, the victory by Frazier was "a proper end to the Sixties. . . . [with] Cassius/Ali belted incredibly off his pedestal by a human hamburger, a man on the verge of death. Joe Frazier, like Nixon, had prevailed for reasons that people like me refused to understand." For Northern Ireland activist Bernadette Devlin, Ali's loss was nothing short of "tragedy." When the journalist Jimmy Breslin told her immediately afterward, "You act like you lost yourself," the forlorn Devlin replied, "I did."

Jerry Izenberg dropped in on Joe at his gym a few weeks after the fight. "He had steam coming out of his ears," the *Newark-Star Ledger* columnist would remember. Ali had been on television giving his assessment of the outcome and, as Izenberg observed, "Only he could convince America that he actually won." On one of the walls of the gym was a blown-up photograph of Ali on his back in the fifteenth round. When Izenberg commented on it, Frazier started to say, "That motherfucker . . ." But he stopped himself and said instead, "Come on, we'll get some sandwiches. Then we'll come back and I'll tell you what I really think." They hopped in the car and drove to a nearby deli, where one of the assistants who had come along went inside for the sandwiches. As Frazier, Izenberg, and another assistant stood outside on the sidewalk, three young black boys came running toward them.

"Joe Frazier! Joe Frazier! Joe Frazier!"

Frazier turned to the other assistant and said, "You gotta go to the car. Get the autographed pictures. Bring them back. Hurry up. I want to help these kids." He then turned to the boys.

"What are you doing out?" he said. "You should be in school."

"It's lunchtime," said one of the boys.

"Jeez," Frazier said, handing them each a picture. "I never thought of that."

One of the boys then said: "My daddy says Muhammad Ali was drugged."

Izenberg would remember, "Joe came as close as he had ever been to turning white." Frazier bent down, looked the boy in the eye, and said, "You run home and you tell your daddy, he's right. He was drugged. I drugged him with a left hook."

CHAPTER EIGHT

DOWN GOES FRAZIER

Joe and Dolly, shelling beans on the porch at Brewton. *Philadelphia Bulletin*

Live Oaks draped with Spanish moss engulfed the weathered house he lived in as a boy. In the front yard were bare patches of dirt and the rusted remains of a broken-down car. Out in the back, where some digging would surely unearth the odd jug of corn liquor that had been buried beneath the hog pen years before and long forgotten, a clutch of chickens pecked at the ground and clucked. Not far from the oak tree where he had hung his do-it-yourself heavy bag, which he had loaded with rags and corncobs and bricks, Joe knelt down on one knee, took aim with a .22-caliber rifle, and began squeezing off rounds at a row of bottles and cans. Mostly,

he missed; Joe was always erratic with guns, once accidently discharging one into the leather upholstery of one of his cars. With an
eye on the laundry she had hanging on a line nearby, Dolly warned
him not to shoot holes in her only set of good sheets.

Twelve years had passed since he had shoved off on a Greyhound bus with a change of clothes and a sack of fried chicken. In
the wake of his victory over Ali, he had come back in April 1971
with Florence and their five children in two Cadillacs, and with
an invitation to speak before the South Carolina State Legislature. He would become the first black male to do so since Reconstruction. In an interview with reporters upon his arrival back in
Beaufort, he said he just planned "to rap with those fellas" at the
statehouse, not deliver a political speech. The previous fall, South
Carolina had overwhelmingly elected a moderate liberal governor,
John C. West, who had promised in his inaugural address that under his administration the government would be color blind, but
Frazier had visited enough through the years to know that whatever change had occurred had been small, and that "some damn
people never gonna change." Were it not for the fact that, as he
said, "I got a mommy down here," he was not certain if he would
ever come back. Seeing her baby boy again, the old woman would
say with a gleam in her eye, "Dolly Frazier, mother of the Champ.
How sweet it is."

Only the date on the calendar had changed in South Carolina.
Philadelphia Daily News columnist Tom Cushman discovered that
when he joined Joe on his trip back home. With a night to kill, he
stopped in a tavern on Parris Island to sop up a beer and absorb
some local color. Having once served in the Marine Corps, he had
heard Parris Island was one of the worst places on earth, given to
unbearable heat and swarms of mosquitos. At the bar was a group of
drill instructors. "We got into a conversation," Cushman told me.
"Once they discovered I had been in the Marine Corps, I had no

trouble talking with them." Upon hearing that he had come down from Philadelphia with Frazier, it quickly became clear to Cushman that they looked upon Joe as "almost a national hero" for beating "Clay," whom they despised for his evasion of the draft. From 1962 until 1973, two hundred thousand recruits passed through Parris Island on their way to Vietnam, while "Clay" hid behind lawyers and shot off his big mouth. These leathernecks loved Frazier. "Now, there is a tough guy," one of them told Cushman. "He gave that nigger the whipping he deserved."

A standing-room-only crowd of five hundred people came to hear Joe speak at the South Carolina State House in Columbia. He had come up from Beaufort in a six-car caravan led by a police escort that morning. In a meet-and-greet, Governor West presented him with a silver jewelry box, as Florence and her five children looked on with pride. Ushered into the House chamber to a thirty-second ovation, he was introduced before the joint session of the legislature by State Senator Ralph Gasque, a champion of underprivileged blacks who, as a former amateur boxer, held the distinction of having leveled a fellow pol with a one-punch knockout in a quarrel on the senate floor. By way of setting the scene, it did not go unnoticed by Cushman, Dave Anderson of the *New York Times*, and other reporters in attendance that the Confederate "Stars and Bars" was displayed along with the American and state flags; and that there were only three blacks among the 124 representatives in attendance and none among the forty-six senators. While Frazier did not write the speech he would give (it was penned by his publicist, Joey Goldstein), it expressed his deepest beliefs in that it called for unity between the races. And he delivered it with an aplomb that would have seemed improbable for someone who just a few years before had been recommended for elocution classes.

"I say, we must save our people, especially the young ones," he said. "There's this drug thing now. Sometimes I'm driving around and I see kids, moving through the streets of Philly, heading for trouble, and I wonder to myself, 'Where's Mom and Dad?'

"I come back to my hometown of Beaufort, and I can see some things have improved. But too many things haven't. Mostly, it has to do with people and the way they think.

"I sit down and ask myself, 'Why does this have to go on, and on, and on?' We don't have time for it. We need to work together, play together, pray together, and do everything together."

To advance the ideals he had spoken of, Joe volunteered to build a playground in Beaufort. According to Joe Hand Sr., who stayed behind in South Carolina to iron out the details, the offer fell apart after an awkward exchange with a city official, who pointedly asked: "So who's going to use this playground?" Once Hand had explained to him that it would be used by children of both races, and that Joe was adamant that there would be one and not two separate water fountains, he received a call in his motel room from a detective with the Beaufort police. Genial yet firm, the man said, "Look, you would be doing me a favor and yourself a favor if you took the next plane out of here." Hand peeked through his window blinds and saw two unmarked cars idling in the parking lot. He packed his bag. A plainclothesman then picked him up and dropped him at the airport. Hand later told me, "They did not want a black person building a playground in a white neighborhood that was going to be racially integrated." Hand flew home to Philadelphia and reported back to Frazier.

"Fuck 'em," Joe snapped. "We won't build it then."

Was Joe surprised?

"Not at all," Hand said. "He was more surprised that I was surprised."

From South Carolina, it was on to Washington and the White House, where President Nixon invited him for Sunday services. Speaking in the East Room before an assembly of three hundred was Reverend Carl W. Haley, of the Arlington Forest Methodist

Church; joining him were the Singing Cadets, a fifty-seven-member choir from Texas A&M University. In introducing Frazier, Nixon observed: "Nothing comes easy in life. Not everyone can be Champ; not everyone can be athletic. But everyone can be the best and try to make something of himself." Frazier and Florence stood in the receiving line with Mr. and Mrs. Nixon. When a reporter asked Frazier for his impressions of the sermon by Haley—the thrust of which was how "compassion is the cardinal principle of progress"—he said it was "beautiful," yet noted: "The reverend was a little long-winded." Florence told another reporter that she hoped Joe would give up boxing, observing: "I want him to be a husband and work like a regular man." Told what she had said, Frazier said with a shrug, "All wives are like that." In a conversation later that day with White House counsel Charles W. Colson, Nixon called Frazier "a fine guy."

> COLSON: Well, he apparently is, uh—
>
> PRESIDENT NIXON: Strong.
>
> COLSON: —from all the reports we got, he is sympathetic to you.
>
> PRESIDENT NIXON: Well, whether [or not], he certainly is good with young people. I mean he—
>
> COLSON: Is that right?
>
> PRESIDENT NIXON: —he had the young blacks, he says, "Look, you've got to work, kids." Well, that's why we had him.
>
> COLSON: Well, we have Sammy Davis, who is—
>
> PRESIDENT: Yeah.
>
> COLSON: —smelling around—
>
> PRESIDENT NIXON: We'll see. Wait just a little while and then we'll do it. [*Laughs*.]
>
> COLSON: If we, if we pick up a few fellows like that, uh, it has quite an effect.

From the South Carolina Legislature to the Nixon White House and on to Philadelphia, it seemed to be raining politicians. On his way back from Washington, he stopped at the Shack Restaurant for a fifty-dollar-a-head fund-raiser for Rizzo, who announced to his nine hundred partisans in attendance: "I couldn't let the city fall into the hands of the lefties." And there he was the following day, atop the back seat of an open convertible with Mayor Tate in a motorcade through Center City in celebration of "Joe Frazier Day," the fourth such event in his honor since he had come back from the Olympics. Upon canceling an appointment the following Tuesday to speak before the Pennsylvania State Legislature in Harrisburg, he turned his attention from political glad-handing to singing. To get his voice back into shape, he hopped back down to Charleston for some shows with the Knockouts at the Porgy and Bess Club. There, he conceded to a reporter that Florence was not pleased with his vocation. "To her, it's one big headache," he said. "She would like me to stay home and be a gardener, or a bellhop around the house." From Charleston, it was off with the Knockouts for a three-day engagement in Lake Tahoe and then to Europe, where in May and June 1971 he had a six-week slate of concerts in twenty-nine cities.

The tour was a flop. Fewer than one thousand fans showed up for their first performance, at the seven-thousand-seat Pellikaan Hall outside Amsterdam. Only 250 people came out to see him in Cologne, and Frazier had to be persuaded to go on by the concert organizer. Five hundred patrons came to see him at the eight-thousand-seat West Berlin Sport Palace, which caused one cynic to observe that "it looked like there were almost as many persons onstage as in the audience." Joe shrugged it off, saying: "It looked like a crowd to me." When his traveling party of twenty-six moved on to Vienna, Joe was nowhere to be found, which caused a fair amount of consternation. What happened to Joe? He failed to show up for a luncheon with U.S. ambassador John P. Humes and a group of Austrian sportswriters. When he showed up the following day, he explained, "Yesterday was my day off." Only three hundred fans came to see him in Vienna at the two-thousand-seat Stadthalle. The show in Copenhagen was canceled when only twenty-eight seats were sold. When he was two hours late for an appointment with Lord Mayor Joseph Cairns in Belfast, Cairns huffed: "I am a very busy man and there is a limit to how long I can hang around." Apparently, Frazier had been stopped in his Rolls-Royce and searched by British troops. At an engagement in Limerick, Frazier walked off the stage when only sixty people showed up, leaving the annoyed theater manager to announce that the show would not go on. Upon his arrival back in the United States, Frazier observed plaintively, "Man, you can't club people over the head to make them come out."

Although he would say that he would "rather sing than fight," there was no getting around the reviews. Even ardent admirer Red Smith once chimed in with a poke upon hearing him perform at the Concord Hotel, observing that the Knockouts "sounded like kitchenware falling down stairs." Soon after Frazier returned from Europe, he played with the Knockouts at the Temple University Music Festival in Ambler, Pennsylvania, in a joint program with the Pittsburgh Pops, and was once again skewered critically. Writing in the *Philadelphia Inquirer*, Samuel L. Singer observed, "There is no more reason to pay to hear the world boxing champion sing than there would be to pay to see Engelbert Humperdinck or Roberta Flack box." Singer recommended that the Knockouts acquire a "coach," lest they be destined to "run out of places to visit for the first and last time." Although Joe said he never read the paper, he began taking voice lessons. Coach Carlo Menotti had him doing exercises that included singing while lying on his back with his legs up in the air. He quipped, "In a month, I'll have him singing 'La donna e mobile.'"

George Foreman would remember years later the fear Joe Frazier instilled in him. Given how the two bouts between them unfolded, it seems possible that he was merely giving a chummy nod to an old friend. But in a *Playboy* interview in 1995, he spoke of how "really scared" he had been of Frazier when they fought in Kingston in January 1973. Along with the fact that he knew Frazier never let up, that he pursued his opponent with the aggression of an unfed wolverine, it did not escape Foreman that Joe had "that look." He had seen it before. It was as hard as steel, and it reminded him of the glower he had seen on the faces of some of the characters he used to tangle with as a strapping young juvenile delinquent in the Houston Fifth Ward, called "The Bloody Fifth" for the shootings and stabbings that regularly occurred there. Although he would betray no such anxiety in the days leading up to his encounter with Joe, assuring the press that he was "gonna knock him stone cold," Foreman would concede years later that Joe was the only opponent who had ever intimidated him. "Nobody else," he said. "Before or after."

Big, strong, and dangerous with either hand, Foreman had endeared himself to the American public at the 1968 Olympic Games in Mexico City. In what would stand as a counterpoint to the provocative protests of sprinters John Carlos and Tommie Smith, who, during the medal ceremony, each famously raised a gloved fist in a salute to black power, Foreman waved a small American flag in the ring upon capturing the gold medal. He had only been boxing a year. With an amateur record of 22-3, the young man who turned his life around by joining the Job Corps at age sixteen found himself at the doorstep of a potentially lucrative pro career. Under the guidance of a trio of veteran boxing hands—Dick Sadler, who had worked with Archie Moore and Sonny Liston; his cousin Sandy Saddler (spelled with two *d*'s), the former world featherweight champion who would be remembered for his four brutal bouts with Willie Pep; and Moore himself, who held the world light heavyweight championship longer than anyone in history (December 1952 through May 1962)—Foreman debuted professionally on the undercard of the Frazier-Quarry bout at Madison Square Garden by knocking out Don Waldheim in June 1969 and clicked off thirty-six consecutive victories. Of the thirty-three knockouts he recorded, thirty-two came in fewer than five rounds, and ten of those came in the first round. But the quality of his opponents was something less than top tier. Only Gregorio Peralta (whom he beat twice) and George Chuvalo were ranked. Still, the WBA positioned Foreman behind only Ali as a contender, conveniently assigning him to the top spot when negotiations for Ali-Frazier II stalled and Frazier agreed to defend his championship against Foreman.

Neither Durham nor Eddie Futch was in favor of giving Foreman an immediate shot. Taller, heavier, and with a longer reach than Frazier, he was the type of opponent who, as Durham would say, "could leave you with your brain shook and your money took." Plus, Ali would bring with him the certainty of a far better payday than the $850,000 Frazier would get for fighting Foreman. Ideally, Durham and Futch would have preferred to see Joe fight Ali a second time and call it quits while he was still in good health. But an obstacle stood in the way. Though Frazier had said again and again that "if Clay is ready tomorrow, I am ready today," nothing happened. Through the press, they sniped at each other. Joe claimed that, since he was the champion, "Clay needs me!" Ali claimed that without *him*, Frazier would not be able to "draw flies." "Joe needs me!" said Ali. According to Frazier, Ali would occasionally call him and say, in a conspiratorial whisper through the receiver: "We two big bad niggers, why don't we fight?" But Ali demanded an even split of the purse, and as far as Frazier was concerned, that was not going to happen. He would die and go to hell before he gave

Ali that. Frazier would later clarify his position, saying: "If he had conducted himself right and respected me as a man, it would be different."

Frazier had come to look upon Ali as something less than "a true brother." Although he once had empathy for him in light of his legal woes with the government, which were cleared up when the Supreme Court upheld his appeal in June 1971, Frazier had grown weary of the ad hominem assault Ali had waged against him and how it uprooted his support in the black community. At the Ohio State Penitentiary in Columbus in November 1971, for a special taping of *The Phil Donahue Show*, he encountered a pro-Ali inmate population that jeered him and asked if it annoyed him to be called "a great white hope." Frazier replied that he was nothing of the kind, and that he objected to "Clay always going around knocking white people." A letter writer to the *Philadelphia Daily News* observed: "Being black, I feel about 90 percent of our people also feel that Joe (Uncle Tom) Frazier is not the champ of the world. . . . Wake up, Joe, you're not white." In an apparent effort to whip Frazier into a fury and get him in the ring again, Ali captured headlines by claiming "white America had given Frazier a crumb." Along with calling Frazier "ugly," disparaging his verbal skills, and pointing out that only "twelve or thirteen people" showed up when he sang, Ali was quoted by Stan Hochman in the *Philadelphia Daily News* as saying: "Man works in a meathouse all his life, he's thankful for a crumb. White folks who lynched niggers and killed niggers, then give a nigger a crumb and expect him to be thankful. White America is hated, even by white Europeans."

Irritably, Frazier replied, "Why does he have to talk that way? Why does he talk hatred and separation?"

Going into the Foreman bout, no one could be sure of the extent of the wear and tear Frazier had suffered in his victory over Ali. Close to two years had passed, and he had fought only two bouts, neither of which lasted long enough to provide any definitive answer. On the evening before Super Bowl VI, between the Dallas Cowboys and the Miami Dolphins in New Orleans, he came off a ten-month layoff for a January 15, 1972, bout with Terry Daniels, a handsome former football player from Southern Methodist University who had a 28-4-1 record. Tipping the scales at his heaviest weight ever—215, twenty-four pounds heavier than the challenger—Frazier floored Daniels five times before the fight was stopped at 1:45 of the fourth round. Wryly, Daniels said later, "They needed a math major out there instead of a referee."

A still-heavier Frazier was back in action four months later in Omaha, Nebraska, where he was paired on May 25 with local favorite Ron Stander, 23-1-1. From across the Missouri River in Council Bluffs, Iowa, Stander was backed by a crowd of admirers wearing straw hats bearing the question, WHO THE HELL IS JOE FRAZIER? The answer to that question would arrive on the end of a left hook. Through four bloody rounds, Frazier—a half pound lighter than Stander at 217½ pounds—carved up the so-called Bluffs Butcher into a deli tray before the fight was halted in the corner before the fifth round. To sew the likable lug back together, it took a half hour and seventeen stitches: eleven on either side of his nose, and six above and below his right eye. Something less of a dreamer than her battered husband, Darlene Stander, his wife and high school sweetheart, said famously of the bloodletting that had just occurred: "I'm a realist. You don't enter a Volkswagen at Indy unless you know a helluva shortcut."

Unbeaten now in twenty-nine consecutive bouts, Frazier appeared off his game during his preparations for Foreman. Family had come down to Jamaica for what had been planned as a reunion, with Florence, the children, his siblings, and others lounging by the hotel pool. In his autobiography, *Inside the Ropes*, referee Arthur Mercante would remember that "the atmosphere [in his camp] was too relaxed, too festive. There were parties, barbecuing and songfests long into the night. Frazier seemed more focused on his second career as a rock singer than preparing for a heavy-weight title fight." Eddie Futch was immediately troubled by what he saw when he arrived in Jamaica and told Durham so.

"Yank, do you know what you've got here?" Futch said. "You've got one big party."

High rollers had come in from New York, Miami, and Los Angeles. "They were all there," said Futch. "They all came down to take their vacations." And Joe joined in. Futch would remember Joe was sitting poolside in an expensive suit when some people in the pool called for him to come in. "Instead of sending someone to get him some trunks, he asked for a pair of scissors," said Futch. "He cut the legs off the expensive suit pants and got into the pool." Later, he asked Futch if he could check out a party. He told Eddie he would be back in ten minutes. When fifteen minutes elapsed and Joe was still not back, Futch drove over to the Kingston night-club where Frazier had said he would be. There, he found Joe up onstage singing with his group, the Knockouts.

Futch was livid. "I went up onstage, grabbed the microphone, and hit him on the head with it," he said. "I got him out of there."

Futch was even more concerned by what Frazier had shown during his sparring sessions with Ken Norton, who two months later would upset Ali in San Diego. Futch remembered, "I was sur-prised by what I saw. It was very even. On the second, I had him spar with Joe [again]. Again, the same thing. By the third day, I called Ken aside and said, 'There's something going on in the ring that I can't fathom. What is going on?'" Norton replied, "He seems to have lost his drive." With only days remaining before the bout, Futch told Norton he was now on "vacation" and yanked him from the rotation of sparring partners, if only to keep Joe from losing confidence in himself. Contrary to what Foreman later said—that he was never more afraid of an opponent—he observed in the days leading up to the bout: "The end is near for Joe Frazier." Of the strategy he planned to employ, he simply said, "I'll just stand there and POP! POP! POP! Just like shootin' birds off a fence."

On an evening that Kingston boxing promoter Lucien Chen hoped would bring "a recognition of the fact that Jamaica is a place where astounding things happen," a crowd of thirty-six thousand pushed through the turnstiles at the National Stadium and were presented with something astounding indeed. With a four-inch edge in height and reach, the six-foot-three Foreman towered over Frazier, their eyes locked in an icy and unflinching stare, as Mercante conducted the prefight instructions in the center of the ring. To equip himself to use his height to his advantage, Foreman had worked assiduously in the weeks leading up to the bout to perfect his right uppercut. For Frazier to be effective, he had to do what he always had: stay down low, get inside, and work the body. But as the first round unfolded, it was clear that Joe was up too high, as Foreman used his superior size to push him off, which he would fourteen times in what Futch called a "fouling tactic." Then—at 1:40 of the round—Foreman stunned Frazier with a left hook, quickly followed up with a left-right combination, and then dropped him with a concussive right uppercut. From ringside, announcer Howard Cosell shrieked, "Down goes Frazier! Down goes Frazier! Down goes Frazier!"—words that would live on in schoolyards across America whenever any foe was subdued, particularly if his surname happened to be "Frazier."

Twelve-year-old Marvis Frazier shouted from ringside, "Get up, Pop!"

Frazier sprang up and waited out the mandatory eight-count. Again, he engaged Foreman, who drove Joe to the ropes with a volley of hard blows and floored him once more with a cruel right uppercut, perhaps the hardest punch Frazier ever absorbed. From ringside, Marvis shouted, "Stop playing, Pop! Get up!" Again, Frazier scrambled to his feet. Again, Mercante counted to eight.

He would remember, "I actually had to restrain Frazier from getting back at Foreman too quickly. This guy was one tough son of a bitch." Seconds before the round ended, Foreman popped yet a third bird off the fence, as it were, when he caught Joe with a sequence of one-two power punches that sent him sagging to the canvas. Since the rules that were in force stated that a downed fighter could not be saved by the bell, Mercante continued his count after the round had ended. Marvis yelled, "Stay down, Pop!" Unsteadily, Frazier regained his footing and staggered back to his corner, where Durham gave him a deep whiff of smelling salts. A roar went up from the crowd as Cosell exclaimed, "Oh, what a first round!" Joe drooped on his stool, his eyes clouded over, as Durham affixed an ice bag to his neck.

"C'mon, Joe, baby!" Angelo Dundee shouted from his seat at ringside. With a figurative wink, Ali had sent Dundee to Kingston under express orders to "make sure nothing happens to Joe Frazier"—and thus preserve the jackpot that awaited them both in their rematch. But as the second round got under way, Big George was on top of him again. Slipping a wild left hook, Foreman staggered Frazier with an overhand right, which buckled his knees. Instinctively, he gravitated to ring center. As Frazier turned away, Foreman caught him over the shoulder with a chopping right hand to the jaw that sent him again sagging to the canvas. Gamely, Joe got up. Down he went again seconds later under the crushing force of a left hand—and again he got up, blood now trickling from his mouth. Mercante observed in his autobiography that "Frazier was undergoing the worst aerial bombardment since *Thirty Seconds over Tokyo*." In keeping with the tradition of giving the champion "every chance," Mercante allowed the drubbing to continue. Foreman seized upon Frazier once again and tagged him with a right uppercut to the chin that lifted him off his feet and dropped him to one knee. Standing over him, Foreman shouted to Durham, "Stop it or I'll kill him." As Joe picked himself off the floor for a sixth time, Mercante stepped in and waved the fight over. The world had a new heavyweight champion.

As Foreman was carried from the ring on the upraised hands of his jubilant fans, Frazier was escorted back to his dressing room by a phalanx of police with riot-control shields. He held his head high. When he spoke with reporters a half hour later, he had a cut on his lower lip that a doctor had sewn back together. "I underestimated him," Joe said. "He was strong. He hit so hard. That first [knockdown], I should have stayed down [low] and protected myself. But I went back at him. I was such a bullhead, you know? I never got it together after that." Someone asked him if there was any truth to the belief by some that the Ali fight had left him depleted. "A lot of people say a lot of things," Joe said with a shrug. "There will be another day." Ali was not so certain. Contacted by a reporter, he scoffed that "Frazier is a dumb, dumb, dumb fighter. . . . If he had been smart, he would have fought me again and then he would have made some big money. . . . He's finished." But in an interview with Hochman back in Philadelphia two weeks later, Frazier conceded that his undoing at the hands of Foreman was a consequence of his own foolish pride. Never one to back up an inch in the ring, he got up off the floor and reengaged Foreman as if his manhood had been challenged in a Philadelphia gym war. Now, Frazier said, there was only one place for him to go: back to the drawing board.

On his way back to the gym and the challenges that awaited him there, Frazier stopped in Rotonda, Florida, to participate in *The Superstars*, a two-day, made-for-television event that showcased nine of the top athletes of the day across ten disciplines, including tennis, golf, swimming, bowling, weight lifting, baseball hitting, table tennis, the one-hundred-yard dash, the half-mile run, and bike racing. Competing along with Frazier were pole vaulter Bob Seagren, Alpine skier Jean-Claude Killy, tennis champion Rod Laver, race car driver Peter Revson, baseball star Johnny Bench, NBA star

Elvin Hayes, NHL standout Rod Gilbert, bowler Jim Stefanich, and NFL legend Johnny Unitas. Each was asked to enter seven of the ten events.

Choosing to take a pass on golf, tennis, and table tennis, Frazier should have found a way to avoid the swimming pool instead. Matched with Revson, Killy, and Seagren in the first heat of the fifty-meter swim, Frazier proved that Dolly knew what she was doing when she warned Joe to stay away from the creek as a boy. At the gun, Frazier splashed into the water in a belly flop and quickly fell to the rear, as ABC Sports reporter Jim McKay announced: "And Joe Frazier is well, well behind." As Revson surged to the lead—he would be the eventual winner in 32.95 seconds—Frazier paddled hard but seemed to get nowhere. He stopped at the wall of the pool, took five deep breaths and continued, his chin craned above the surface. As he exited the pool and grabbed a towel, the crowd stood and applauded. Wryly, he later observed: "The water hit back."

Of the other sports he entered, he did not do especially well in any, finishing near or in the back of the pack in baseball hitting, bowling, the hundred-yard dash, and the bike race. Given his upper-body strength, it seemed likely that he would clean up in the weight lifting competition. But somehow he came in second to Seagren, who won the competition with a 170-pound lift. To even his own surprise, Frazier could not surpass or even equal that. Using only his arms instead of his whole body, he gripped the bar and lurched forward before letting the weight clang to the ground. "I feel sorry for Joe," said Seagren, who had some training in weight lifting with decathletes at Southern Cal. "He was fighting himself just taking the bar off. If someone had given him some instruction, he would have lifted 240." Overall, Frazier finished tied for last with Unitas and collected thirty-six hundred dollars in prize money. The winner was Seagren, who took home a check for $39,700 ($208,968 in 2018 dollars).

"I think I'm gonna retire," Frazier joked. "This ain't my year." And it would only get worse.

━━━

Spread across 366 acres in Beaufort County, Brewton Plantation was once worked by black slaves whose ownership was passed down through the generations along with the deed to the property. Originally part of a two-thousand-acre royal grant that was bestowed by King George II upon Landgrave Edmund Bellinger and Captain John Bull in 1732, it was handed down in 1771 to Mary Izard Brewton, who was lost at sea four years later with her husband Miles and three children while sailing from Charleston to Philadelphia. Burned down a century later by General Sherman on his March to the Sea, it was later rebuilt and passed through the hands of a succession of owners. In 1930 it was bought by John R. Todd, the principal builder of Rockefeller Center. Out-of-state investors purchased the estate upon his death in 1945 and allowed it to go to seed in the years before Cloverlay lawyer Bruce Wright acquired it in May 1971 as an investment for Joe, who approved the purchase without first seeing it. When he saw the condition it was in, he was furious with Wright. Windows were broken, fences were down, the fields were overgrown, and the ponds were choked with weeds. But when he calmed down, he decided, "Hell, I've worked worse land than this." And so he rolled up his sleeves.

Only nineteen miles from his boyhood home in Laurel Bay, the plantation seemed to be on the other side of the world to Dolly, who was set in her ways and had a particular attachment to the house that her husband had built for them. But she took her "critters" with her, a few hogs, some chickens, and a goat called Billy; planted rows of vegetables out back; and saw to her daily chores, her daughters by her side and grandchildren underfoot. Every day, she would pack her pipe with Carter Hall tobacco and sit on the long porch with her Bible, which granddaughter Dannette helped her to read, or sing to herself a gospel favorite, "Jesus, Keep Me Near the Cross." Whenever Joe showed up—usually four times a year—he would find himself shelling beans with her on the porch or doing odd jobs around the place, such as scraping old paint, repairing fences, and clearing land. Surveying the expanse of his acreage on a tractor, he envisioned one day buying a herd of cattle. Joking, he said he would sell it to Cross Brothers, the slaughterhouse where he had worked as a young man in Philadelphia, if only to show them how far he had come since they dropped him from the payroll when he came back from the Olympics with a cast on his hand.

Behind the wheel of one of his Cadillacs, Joe would roll up to "the big house" with members of the Knockouts in tow and droves of other people at his heels. "It was always party time when Uncle Billy came around," said his niece Dannette. "You know that expression, 'What happens in Vegas, stays in Vegas?' Well, what happened at the plantation . . ." Seven months before he fell to Foreman, he held a Fourth of July party that lasted a full week as hundreds of relatives, friends, and curious Beaufortonians poured onto the sprawling property to rub elbows with the Champ. As Les Pelemon remembered, long outdoor tables were crowded with all sorts of "down-home cookin'"—barbecue, fried chicken, fish, and ribs, accompanied by sides of corn on the cob, potato salad, and greens. Large quantities of beer and liquor flowed liberally from an open bar, including jugs of moonshine. "Joe knew where to go in the county to get the good stuff," said Pelemon. None of this sat well with Dolly, who Dannette said would give her youngest son an earful.

"You think I care about Joe Frazier this, Joe Frazier that?" Dolly would say, her voice rising. "You think I care about all these people? Billy, I want to know if you love the Lord. I want to know: Are you going to get into the Kingdom of Heaven?"

"Mama," Joe would reply. "I bought you this house. . . ."

"Do you think I care about this house? Do you think I care about another hat or shoes? Billy, I want to know if you love the Lord."

"Oh, Mama," Joe would say, tears welling in his eyes. "I work so hard. . . ."

"Do you think I care about money? You can keep your money!"

Dannette laughed as she retold that exchange years later. "Lord, that happened all the time," she said. "All she wanted to know was, 'Are you in good standing with the Lord?' That woman was never one to cuss, but she sure could fuss. And she did not like having all those people around. Of course, if Uncle Billy was bringing them down to be saved, that would have been another story."

Though he was no longer champion, Joe remained in demand in the immediate aftermath of what he called his "mishap" at the hands of Foreman. Twice crisscrossing the country, he showed up at banquets, judged two beauty contests, gave a lecture at Millersville State College, and appeared on *Today* and other television talk shows. He attended the Ali-Norton bout in March in San Diego, where his erstwhile sparring partner upset Ali and broke his jaw; he boxed an exhibition in Seattle; and he dropped off his Lincoln Continental in Detroit for a paint job. Publicly, he professed an eagerness to get back into the ring with Foreman. Although Durham had a handshake agreement with Sadler for a quick return bout, neither he nor Futch was particularly dismayed when Foreman found himself caught up in litigation with a Houston promoter, Ludene Gilliam, who claimed to have an agreement with him to present his first fight as defending champion. Instead, Durham worked out a date with Hungarian-born Joe Bugner, the British and European heavyweight champion who had acquitted himself well in losing a twelve-round decision to Ali that February. Frazier declared that he had recommitted himself to the rigors of training.

A crowd of some fifteen thousand jammed into the Earls Court Exhibition Centre in London on July 2, 1973. Again, Frazier found himself at an anatomical disadvantage. At six foot four and 221 pounds, Bugner was four and a half inches taller and thirteen pounds heavier than Joe, who had trimmed down to his lowest weight (208) since the Ali bout. With the grim perseverance of an axman with a dull blade, he chopped down Bugner across twelve rounds with an unyielding array of body shots, which Bugner would say years later "saw me pissing blood for a week. . . . He would just bore into your body and tear you to pieces. I had serious kidney and liver problems afterwards." Frazier cornered Bugner in the tenth round and stunned him with a left hook to the side of the face. For a second or more, Bugner seemed to hang there as he began to sag down the padded ring post to the canvas. But Frazier did not step in with what surely would have been the finishing blow, which enabled Bugner to beat the count and come back later in the round and stun him with a right hand. Although referee Harry Gibbs would score the bout 6–3–3 in favor of Frazier and thus award him the decision, the talk later centered on the inability of Frazier to vanquish Bugner when the opportunity presented itself in the tenth round. With his left eye swollen shut, Frazier later said he expected Gibbs to stop it. Of Bugner, he told reporters, "The guy was really helpless, believe me."

He would soon be thirty. Talk of retirement had grown louder in the press, which had wondered in the wake of the Bugner bout if Frazier still possessed his "killer instinct." But Joe wavered when the question of his exit came up. Early in his career, he seemed to have a clear picture of when he would get out. He said prior to his victory over Quarry in June 1969 that he would quit after "five more fights." He had said then, "Maybe four and a half years. I'll be twenty-nine. That's a good time." But as he found himself astride that age, he would not allow himself to be pinned down. With the exception of the Foreman bout, he cleaved to the guidance of Durham, who, with Futch, had not just chosen his bouts carefully but had spaced them out with an eye toward preserving his health. While others were beginning to see an incremental erosion in his ability—Futch had observed that he was "a split-second late" in his exchanges with Bugner—Joe still loved the physicality of the sport, the attention it garnered him, and, as he would say, yeah, "the money paid the rent." But no one could predict what the years or even days ahead would bring, as he would be reminded again in August 1973 as he sat playing cards at the plantation and received a phone call from Philadelphia. Someone told him, prematurely, that Yank was dead.

"We were all sitting around playing Rummy 500 or something," said Lisa Coakley, his niece. "When the call came in, Uncle Billy was devastated and distraught. I remember him saying, 'No way Yank can be dead. I was just with him the other day.' He flew out the door, got on his motorcycle, and jumped on a plane to Philadelphia."

Durham had collapsed with a massive stroke and was on life support. Only days before, Joe had been with him in New Mexico as he prepared Bob Foster for his bout against Pierre Fourie. As Frazier remembered later, Durham had a problem with high blood pressure. He had been hospitalized with it in May 1971, which coincided with the diagnosis of one of his four children with sickle cell anemia. When Frazier arrived in Albuquerque, he said Durham sent him out for Bufferin. "He was always using it," Joe said. "When I would go in his suitcase for something, there would always be two or three bottles of the stuff." The last time Joe saw him was when Durham dropped him off at the airport and said he would see him back at the gym on North Broad Street. He appeared fine. Hours before he collapsed on Tuesday morning, August 28, he had complained of a headache.

Joe sat with Yank by his bedside at Temple University Hospital. He rubbed his hands and shoulders, "trying to put the warm touch onto him," as he later explained to Stan Hochman. But the hands were cold and the strong shoulders he had once climbed upon in search of a place in the world were now limp. Gazing upon the tangle of tubes that covered Yank, who had undergone emergency surgery to relieve the pressure on his brain, Joe picked through their eleven years together, which now seemed to have ended. More than a boxing manager and trainer, Durham had become at once a father figure to Joe and a repository of his trust. In a profession teeming with connivers, he was a shield that repelled any deal that was less than top dollar and protected Joe from potential trouble, be it a strange face that suddenly showed up in the gym or a plate of uninspected food in some restaurant. There were no secrets between them. Even when it came to the safe-deposit box they shared for what Denise Menz called "all the scams they had going on the side," Joe had never asked for the number on the account, only because he could not envision a day when Yank would not be around to open it.

Nothing could be done for Yank. Three days after he fell ill, he slipped away at age fifty-two. The funeral was held a week later at the Bright Hope Baptist Church, where would-be fighter turned tailor Billy Johnson joined in the crowd of mourners on the pavement out front. He appreciated the favor Durham had done for him by giving an honest appraisal of his ability and not leading him on. Johnson told *Philadelphia Bulletin* columnist Claude Lewis, "He could have milked me for a few pro fights and made some money." Inside the church, Joe sat in the front pew and wiped away tears, thinking back to how Yank used to give him hell. He would even challenge him to "lace on the gloves," only to reconsider as they approached the ring and say with a twinkle in his eye, "We'll do it another day, now let's talk." Joe could be the same way with his own son Marvis—harsh yet loving.

Immediately after Yank Durham's death, Joe was so inconsolable that observers speculated that he would retire. But with the passing of a few days, he announced his plans to hand over the reins of his career to Eddie Futch, who had been with him since the Memphis Al Jones bout in Los Angeles in 1966. He had come to value Futch for the depth of his knowledge, his quiet yet pointed manner, and his scrupulous attention to detail. With just over a year remaining on his second extension with Cloverlay, Joe looked forward to avenging his loss to Foreman and winning back his championship. Were he to do that, it seemed likely that he would give Ali that return bout and then retire, win or lose. No one was better equipped to navigate the end chapter of his career than Futch, who had what he called a "firm relationship" with Yank. From the day they began working together, Futch found Yank to be astute, jovial, and a pleasure to work for. Aware that his health was spotty, Yank had pulled Eddie aside a month before he died and told him in that gravelly voice: "If something ever happens to me, I want you to take care of my boy."

Five days before Joe and Ali were scheduled to face each other on January 28, 1974, Denise Menz stood in the lobby of 30 Rockefeller Plaza and looked out at the passersby bundled in winter coats and hats as she waited for her ride. For the better part of the evening, she had been up in the Rainbow Room on the sixty-fifth floor going over the menu, guest list, security, and other details for the party Cloverlay planned to hold after the fight, which would have the trappings of a celebratory event regardless of the outcome. Spotting the burgundy Cadillac rolling up to the curb, she hurried outside and came upon her reflection in the tinted rear window. The door clicked and opened from the inside.

Joe scooted over in the back seat to give her room as she slid inside. Up front, Eddie Futch sat in the passenger seat as the driver wheeled the car across midtown, through the Lincoln Tunnel, and onto the New Jersey Turnpike. As Denise cuddled up to Joe and reported on her day, it occurred to her that he was oddly withdrawn.

When he spoke, his words were clipped, not in a way that she perceived as angry or annoyed, but weary. Otherwise, he just stared out the window at the passing lights. Neither Futch nor the driver said much, either. Not until the following morning, when the phones in the gym began ringing with calls from reporters and others, would Denise realize what had happened in the hours before Joe picked her up and what accounted for the mood in the car: At a taping over at ABC with Howard Cosell for *Wide World of Sports* that was scheduled to air that Saturday, Joe and Ali got into a scuffle that ended with them wrestling on the floor.

"What in the world happened yesterday?" Denise asked.

Joe shrugged and replied, "We kind of got into it. I had enough of his shit. Don't worry about it, baby. I've got something for him Monday night."

To preview what was being billed as Super Fight II, Cosell wanted to have Joe and Ali in the studio together and go through the first fight round by round, with each providing commentary. Publicist Bob Goodman remembered that when he approached Frazier with the idea, Joe was "a little reluctant." Futch told Goodman, "Look, Bob. You know Ali as well as we do. Ali is going to keep jabbing at Joe." Goodman agreed that Ali would undoubtedly try to "ignite the situation," both as a promotional ploy and in an effort to get his opponent worked up. Goodman had seen it before, how "Ali could turn it on and off," pretend he was angry and "not care a bit." Frazier was not that way. Goodman observed, "If Joe got upset, he was upset." Goodman turned down the request.

Cosell would not take no for an answer. He called Goodman back and, with Futch on the extension, "guaranteed" them that he would keep control of the show, and "nothing would get out of hand." With that assurance, Futch said, "Okay, we'll do it." According to Goodman, Cosell promised them that he would sit between Joe and Ali. But when they arrived at the studio, Goodman saw that Cosell had the chair on the end of the dais and Joe and Ali were seated side by side. Goodman pulled Cosell aside and said, "Hey, Howard. This is not what we agreed to." Cosell told Goodman not to worry. "Everything will be all right," he said. "I'll be right here." For all the good that did, he could just as well have been poolside back in Jamaica sipping a cocktail with a tiny umbrella in it.

Almost immediately, Ali zeroed in on Joe. Clad in a three-piece brown pin-striped suit, Ali claimed he had been denied victory by a "racist" panel of judges, one of whom had awarded Frazier eleven rounds on his scorecard. Only "white bigots" believed that he lost, Ali opined; "all the black people know I won." Evenly, Frazier replied that he had "won fair and square." As Cosell tried to steer the conversation to the action on the screen, Ali told Joe that the black fans did not like him because he had "too many white followers." Wearing a casual suede ensemble with an open collar, Joe said, "The world is made up of a variety of people." Again, Cosell intervened and asked the two to avoid personal jibes. But the sniping continued and, as Goodman remembered, it was even worse during the commercial breaks between rounds. "Off the air, Ali kept digging at Joe and calling him ignorant," said Goodman. By the tenth round, the wheels came off the cart as Frazier wandered into an area that both agreed would be off-limits: who sent who to the hospital and for how long. Ali became quickly annoyed.

"I went to the hospital for ten minutes," Ali snapped. "You went for a month. Now, be quiet."

Grinning yet agitated himself, Frazier replied, "I was resting. In and out."

"That shows how dumb you are," Ali said. "People don't go to a hospital to rest. See how ignorant you are?"

"Why do you think I'm ignorant?" Frazier seethed. "I'm tired of you calling me ignorant all the time. I'm not ignorant."

Joe unhooked his earpiece, rose from his seat, and stood over Ali.

"Why do you think I'm ignorant?" Frazier said. "Stand up, man."

"Sit down, Joe," Ali said, as his brother Rahman stepped onto the set. Joe turned to him and said, "You in this, too?" Futch came up behind Frazier and reached out to pull him away, but Joe brushed his hand away. Again, Ali said, "Sit down, Joe," only now he was in motion, leaping out of his chair and grabbing Frazier by the back of his neck. From the carpeted platform, they rolled onto the concrete floor, where they ended up beneath a pile of arms and legs as Frazier was joined by his brother Tommy, who went after Rahman. Futch shouted, "Joe, don't! What are you doing?" As the scuffle unfolded, Cosell reported the action from his seat on the set, uncertain if the scene before him was a promotional exercise or actually "real." He decided that Ali was "probably clowning but there is no question in my mind that Joe Frazier is not clowning."

Goodman jumped into the fray in an effort to pull the two apart. "Joe had Ali by the foot and was twisting it," he said. "I tried to pry his hand loose. I said, 'Joe! Joe! There won't be any fight! You'll bust his ankle!' Finally, they broke it up and Joe stormed out of the studio. He was so upset. But Ali was unfazed. He got up off the floor, looked around, and said, 'Give me my comb.'"

Cosell was apologetic on the air. Off it, Futch unleashed his vitriol upon him. "Oh, God, yes," said Goodman. "He hammered him. He told Cosell that he had double-crossed him." Futch told Cosell in parting, "You are an unprincipled man."

One thousand and fifty-six days would separate the actual sanctioned fights between Joe and Ali. From deep in the Pocono Mountains at his rural training camp in Deer Lake, Pennsylvania, where he secluded himself among log bunkhouses and big boulders with the names of the great heavyweight champions painted on them, Ali chopped wood, pumped water from a well, and ate food prepared by his aunt Coretta. Since Frazier had beaten him, Ali had fought a total of 139 rounds across thirteen bouts, losing only a split decision to Ken Norton that he quickly avenged. Though Norton had broken his jaw in their first fight, Ali claimed that he would be "78 percent" of his former self when he tangled with Frazier again, although it was unclear how he had landed on 78 and not 77 or 79. Even as he insisted that "the white man" had robbed him of the victory he claimed he had earned, Ali conceded that he had given away rounds in the first fight by clowning. There would be none of that in their second fight. "No more poems, no more tassels on the shoes," said Ali. "Nothing but boxing." Trainer Angelo Dundee said, "We're working on speed. Speed is what's gonna lick Joe Frazier."

From his gym on North Broad Street in Philadelphia, the interior of which he had done over with the help of Denise and pronounced to be "nice and homey," Frazier dispelled any concerns that he was less than sound. When Futch took over for Durham, he insisted that Frazier undergo a thorough physical exam, which Eddie said he passed with "flying colors." Evidence of his stamina could be found in his sparring sessions, which remained as combative as any main event. "He goes six tough rounds and is not breathing hard enough to blow out a candle," said Futch. With just twenty-two rounds in four bouts under his belt since his victory over Ali, Frazier also pledged to get down to business in his preparation for their rematch, forgoing his excursions with the Knockouts and keeping to himself in a downtown apartment. On January 12, Frazier celebrated his thirtieth birthday—Ali would turn thirty-two five days later—yet Futch seemed to think that the intervening years between the two fights would weigh heavier on Ali because of his swashbuckling style. "I have been watching Ali for years and got a good look at him against Norton last fall," he said. "The legs are gone."

Neither Joe nor Ali was a champion now, yet the fight held an irresistible public appeal, given the spectacular savagery that had unfolded in their first meeting. With the option he held for the rematch, promoter Jack Kent Cooke had hoped to bring it to his arena in Inglewood, California—the Forum—only to encounter hard pushback from Durham whenever the subject came up, swearing he would never agree to fight again in California. (Although Durham would say it was because of "personal reasons," it was rumored that Frazier could not pass the eye exam in California, which Frazier himself disputed.) Ultimately, New York prevailed in the bidding for the event, but not before it offered a more digestible state tax bite and refunded Joe and Ali some three hundred thousand dollars each that had been withheld from their purses from the first fight. Now that both had been shorn of their apparent invincibility—Frazier by Foreman, Ali by Frazier and Norton—few expected their second encounter to generate the same level of intensity that was produced by the Fight of the Century. Nor was the guarantee for Joe and Ali as lucrative: $850,000 plus 33.5 percent of the net receipts. Still, it remained a big New York moment, the same way it always was when Sinatra swung through town and sailed onstage. Moreover, there were some hard feelings in the wake of the scuffle in the TV studio, for which both Ali and Frazier were fined five thousand dollars by the New York State Athletic Commission for behavior "demeaning to the sport of boxing." Ali said, "I can see two guys down in Waycross, Georgia, one of them saying, 'Ernie, we got to see this fight now.'" Frazier said he would have acrimony in his heart when he stepped into the ring.

Even if the energy in Madison Square Garden was at a somewhat lower voltage than it had been on that extraordinary evening in March 1971, the arena was jammed with a capacity crowd of 20,748 for a live gate of $1,053,688; millions of others poured into closed-circuit venues around the world. From the opening bell, Ali was up on his toes and dancing, just as he had done in round 1 of their first fight. Only now, he was moving counterclockwise, an unorthodox tactic in that it seemed to invite Frazier to clobber him with that left hook. One had to wonder: Had Ali been informed by his "spies" or some other source that Joe had challenged vision in his left eye? At the outset, Frazier abandoned his customary bob-and-weave and stood upright, his gloves at eye level. As Norton had done, Frazier dragged his right foot as he followed Ali, which would presumably enhance his ability to plant his feet so he could get his punches off more quickly. Keeping the action in the center of the ring, Ali picked Joe apart with left jabs. When Joe rushed him, Ali grabbed him and held. As the bout unfolded, he employed this tactic again and again.

Ali was again on his toes in the second round, still moving counterclockwise. When Frazier drove him to the ropes and scored with a solid left to the jaw, Ali countered with a sharp four-punch combination, which he followed up with more grabbing and holding. Near the end of the round, Frazier forced Ali out of his counterclockwise pattern with a hook, only to have Ali pivot and unleash two quick jabs. Frazier slipped them, but Ali stepped in with a right hand to the jaw. The blow staggered Frazier, turning him halfway around. As Frazier backed up, Ali let loose a volley of heavy punches, many of which missed or were blocked. As Ali pursued Frazier into a neutral corner, referee Tony Perez separated them and signaled the end of the round. However, the ring lights in each corner had not flashed on. When Perez realized his gaffe, he waved both fighters once again to the center of the ring with twenty seconds remaining in the round. As Frazier moved forward, bobbing and weaving now, Ali tagged him with four solid punches, including a hard left hook to the head at the bell.

Rounds 3, 4, and 5 saw Joe pick up the tempo. Early in the third, he cornered Ali and scored with two hooks to the head, the second of which prompted Ali to shake his head. By now, the pattern of engagement had been established: as Ali continued to circle counterclockwise, Frazier uncharacteristically used his jab in an effort to close the distance between them, whereupon he would once again be wrapped up by Ali in a clinch. Although Ali began to show signs of slowing down in the fourth round, he spun out of a trap Frazier set for him on the ropes, then grabbed and held on. From his corner, Ali overheard his assistant trainer Bundini

Brown wail again and again, "All night long! All night long! All night long!" By the fifth round, Ali had come down off his toes and engaged Frazier with heavier exchanges in the center of the ring. Frazier pressed Ali to the ropes and stunned him with a left hook to the head, at which point Ali smothered Joe again in a clinch. Frazier began to show some swelling over his left eye.

Going into the sixth round, it was clear that Ali was in far better shape than he had been in their previous encounter. With the exception of a brief unveiling of the Ali Shuffle in the first round, he eschewed the clowning that had corrupted that earlier performance and set himself to the sober business at hand, just as he had promised he would do in the weeks leading up to the bout. By tying up Joe in clinches, he was able to offset Joe's aggression, which came now only in spurts and lacked his old urgency. While still highly skilled and giving a worthy accounting of himself, he did not bring the same fury with him to the ring, nor did he come at Ali with the same volume of punches. Ahead in the scoring, Ali outpunched Frazier by three to one in their exchanges. Frazier trapped Ali in his corner and began talking to him, only to have Ali clip him with a sharp one-two to the mouth. Ali then again smothered him in a clinch.

Perhaps knowing he was behind at the halfway point of the twelve-round bout, Frazier advanced from the corner for the seventh round intent upon unleashing mayhem. When he caught Ali with a big left hook to the head, a roar came up from the crowd. Sensing some surge of energy building among the fans, Cosell reported from ringside, "They expect dynamite out of every left hook that Frazier scores with!" As Ali continued to try to wrap him up, Frazier began scoring with regularity now to the body and won the round on all three scorecards. Frazier shot out of his corner and poured it on again in the eighth round as Ali appeared to noticeably tire. He steered Ali into the ropes and pounded him with short shots to the body. Nearing the end of the round, Frazier hammered

Ali with a straight right hand to the head, nearly jarring his mouth-piece loose. At the end of the round, Ali walked wearily back to his corner. Again, Joe had won the round unanimously.

Clearly now back in contention and knowing it, Frazier emerged from his corner even before the bell called him out for the ninth round, his face opened up into a broad grin. He beckoned Ali forward, waving to him with his left glove. Perez pushed him back toward his corner. When the bell did sound, Joe eagerly engaged Ali, landing a solid right to the head before Ali folded him again in a clinch. No longer moving, Ali stood with his feet planted on the floor, as if waiting for Joe to charge him. When he did, Ali pummeled him with fourteen consecutive punches, battering him with straight right and left hands, and short hooks. Wild cheers from the now-standing crowd urged Ali on as he overwhelmed Joe with his hand speed and slashed him with chopping rights and lefts. For Ali, it was perhaps his top round of the evening.

With three rounds to go, Ali was up 5–3–1 on two scorecards and 6–3 on the other. Standing toe-to-toe with Frazier in the tenth round, Ali peppered him with short rights to the head, as Frazier swung and missed. By now, the ridge above his left eye was swollen. Frazier began the eleventh round aggressively, but again, he swung and missed as Ali draped his arms over him in a clinch. Ali was in command now, having his way as he pounded Frazier with sharp jabs to the face. Ali swept all three cards. Knowing that the only way he could now win was by knockout, Frazier came out for the twelfth and final round throwing leather. Wild with his head shots, he dug in with body shots and drove Ali onto the ropes. Though physically drained, Ali spun to the center of the ring and danced away, circling Joe in one direction and then the other. From long range, Ali slapped at Joe with a flurry of punches. Frazier trailed him but was unable to connect with any clean shots. The bout ended with the two tangled in yet another clinch along the ropes.

Ali dragged himself back to his corner.

Frazier briefly raised his arms in victory, yet there appeared to be no joy in his eyes, as if he expected the verdict that announcer Joe Bostic would deliver amid the crowd that had formed in the center of the ring: "The winner, by unanimous decision, is Muhammad Ali!"

For Ali, it had been a clever tactical victory. Intent upon dropping bombs on Joe and taking him out inside six rounds in their first fight, Ali had been content in their rematch to box Frazier, pile up points, and slip away with a decision. Goodman said, "Ali fought the fight he thought he had to fight to win. He would go bip, bip, bip and grab. He would get Joe in a clinch, and it was like a death grip he had on him." There were 133 clinches in the bout. Futch excoriated Perez for allowing Ali to grab Frazier by the back of the neck and push him down. Futch would later say that while he and Frazier had worked on a "tactic" to offset that, Joe was "so emotionally high" due to the scuffle he'd had with Ali in the television studio that he overlooked it. Goodman agreed that the incident "really threw Joe off." He added, "Boxers have to remain in focus. They have to keep a cool head. Once that happened with Ali, Joe was trying to punch holes in the gym walls every day. Ali knew exactly what buttons to push with him."

Upon hearing the verdict, Cosell turned to Frazier amid the commotion in the ring. "I got no argument about nothing," said Frazier, swaddled in a white robe as sweat beaded on his swollen face. Of the dozens of questions that would later be posed to him by reporters, only one had any bearing on what remained of his career, and Cosell asked it: "What happens now, Joe?"

CHAPTER NINE

MAY POPS

Joe and Eddie Futch, 1975. Getty Images

lose to a quarter of a century had elapsed since Joe Louis
passed through Beaufort on his exhibition tour in March 1950.
Though he remained a revered American icon, the years had
been hard on him. Under the crushing weight of Internal Revenue
Service obligations, he sold himself cheaply as a villain on the pro
wrestling circuit, where he squeezed his now-flabby physique into
tights and turned a deaf ear to the laughing crowds. With a trail of
failed businesses behind him, he collapsed on a New York street in
1969 in what was believed to be a physical breakdown. Only later
would it be revealed that he had developed a cocaine addiction,

which led to his admittance to a psychiatric hospital for episodes of paranoia. With his own inconceivable destiny still far away, Ali looked upon Joe Louis with staggering hubris, citing him as an example of how he himself would never wind up. Now sixty, the Brown Bomber had worked as a greeter at a Las Vegas casino and scraped together work where he could find it. At Madison Square Garden on June 17, 1974, he found himself in the same ring with the man who, years before, as young Billy Frazier, had looked upon him with such wonder.

Only now Louis was a referee, hired by Garden boxing president Teddy Brenner to officiate the rematch between Frazier and Jerry Quarry as an added attraction. But the press was skeptical. Along with the fact that he was getting on in years, he had largely refereed pro wrestling matches, which in no way compared to the unscripted warfare that could be expected to erupt between Joe and Jerry. Five years after their previous bloodbath, the stakes were even higher: both were at the crossroads of their careers. The winner would get a shot at whoever prevailed in the Ali-Foreman showdown scheduled for the fall in Africa, and the loser would be more or less through. Would Louis be able to keep order as the action intensified? Or would he find himself overwhelmed by it and be a step slow to react, just as aging referee Ruby Goldstein had been at the old Garden in March 1962 when Emile Griffith backed welterweight champion Benny Paret into his corner and pummeled him with a barrage of unanswered blows? Paret died in a Manhattan hospital ten days later of a brain hemorrhage.

With beads of sweat still pouring off him moments after his loss to Ali, Frazier had an unequivocal answer when Howard Cosell asked him, "What happens now, Joe?" There was no question that he would fight again. Retirement had always been a moving target for him, and now it seemed to be slipping further away, even as there seemed to be concern behind the scenes regarding his health. Three weeks after his loss to Ali, he visited his ophthalmologist, Dr. Myron Yanoff, and complained of worsening vision in his left eye. Upon completing his examination, Yanoff reported to Frazier that his while his cataract condition had progressed only "somewhat" in his right eye, it had advanced "quite significantly" in his left eye. Dr. Yanoff remembered, "I told him that continued fighting would cause the cataracts to worsen." Even so, the condition of the eye worried Futch less at that point than the lack of intensity Frazier had shown in the ring. Futch and Frazier agreed that if he did not beat Quarry, and do so convincingly, it would be time to call it quits. Futch told the press, "He has to show me he still belongs in this rough business."

Annoyed by the scuttlebutt that he no longer possessed the will to whip himself into shape, Frazier did precisely that during seven weeks of hard preparation for Quarry at his North Broad Street gym. Lending Futch a hand now was George Benton, the artful middleweight contender whose career had ended when he caught a bullet in his back in a random shooting incident. Along with the skull sessions they held with Frazier, Futch and Benton tutored him in how to shorten up his left hook and position himself to use both hands instead of just his vaunted left. To drive the point home, Futch yelled from ringside, "Use your right! Use your right!" until he was sure Joe was hearing it in his sleep. Chiefly with that right hand, Frazier tore up his sparring partners as if they had all taken money off him shooting craps. Contrary to the preposterous rumor that circulated in North Philadelphia that Moleman Williams floored Joe in a sparring session—which Futch said would not have happened even if he had handed him a bat and given him a free swing—Frazier actually dropped *him* with such punishing force that Futch was unable to use him again for a week. Next, Frazier toppled the cagey Jimmy Young, left former opponent Scrap Iron Johnson with bruised ribs, and then drilled Frankie Steele so hard that he drove his teeth through his lower lip, a cut that required six stitches to close and nine days to heal. Futch observed, "I have never seen Joe punch any harder or look any better." Across a total of ninety-four rounds of sparring, it seemed as if the "old" Joe Frazier had reemerged.

Quarry remained an enigma even to himself. As he once said, he always seemed to fight the wrong fight, brawling when he should have been counterpunching and counterpunching when he should have been brawling. Although Jerry had earned in excess of a million dollars in the ring, more than any boxer who had not won a championship, he could never find a way to win when the chips were down and greatness called. Along with his losses to Jimmy Ellis and Frazier, he was stopped twice by Ali in bouts that could have positioned him among the elite of the division instead of a step just below. On the heels of his second technical knockout at the hands of Ali, yet another thorough beating in which Ali summoned the referee to stop the fight seconds into the seventh round, Quarry announced his retirement to pursue a career in public relations. By way of an explanation, he told the Associated Press that his "desire" had dulled and the PR job had been just what he had been looking for.

Only six months elapsed before Jerry unretired. In the year and a half that followed, he hired a new manager, divorced and remarried, and clicked off six consecutive victories, including a unanimous twelve-round decision over previously unbeaten ex-con Ron Lyle and a first-round technical knockout over hard-hitting Earnie Shavers. Unveiled before the boxing press prior to the Frazier bout would be a "new" Jerry Quarry, free now of the corrosive influence of his father and the managerial entanglements that had accompanied him. New manager Gil Clancy, who had shrewdly pointed the way for former welterweight and middleweight champion Emile Griffith, enthused that "Jerry is on the way up." According to Clancy, he had become physically stronger, a harder worker, and less sensitive to criticism. To go with his repackaged look inside the ring, Quarry also had a new wife, Arlene Charles—

called "Charlie"—a blond former Miss Indiana and budding starlet whose credits included parts in *Clambake*, *Speedway*, and *I Sailed to Tahiti with an All Girl Crew*. Charlie explained that she was around to keep Jerry "relaxed." Concerned that his fighter would climb into the ring too relaxed, Clancy warned Jerry to lay off sex for the two weeks before the bout.

Odds dipped from 6–5 Frazier to even money with the arrival of fight night, clear indication that regardless of how Frazier had drubbed him in their prior meeting, the personable white heavyweight had a loyal following of fans who had warmed to reports of his reclamation. But Frazier was certain that the new Jerry Quarry would quickly revert to the old one as soon as he laid leather on him, and that is exactly what happened. Only in the first round and briefly in the second did Quarry try to box Joe as Clancy had instructed. From that point on, it was all Joe. Grinning as he advanced upon Quarry, he pounded him to the body and head, as Quarry began bleeding from a cut over his right eye. Frazier worked Quarry with both hands to the body in the third and stunned him with a solid right cross to the head. Clancy yelled, "*Mooooove! Mooooove!*" Quarry stood as if his boxing shoes were encased in concrete while Frazier pummeled him again and again. Through the fourth round, Frazier continued to pour it on, backing Quarry into one set of ropes and then another. With thirty seconds remaining in the round, Frazier landed a double left hook to the head and chin, which appeared to freeze Jerry in midair before he caught himself from falling. Just before the bell, Frazier connected with a fierce left hook to the abdomen that caused Quarry to sink to the canvas. Quarry got up at the bell as the count reached five. With tears in her eyes, Charlie left her seat and headed toward the dressing rooms.

Standing over him before the fifth round, Clancy alternately scolded Quarry for "throwing the fight plan out the window" and worked to close the cuts that had opened up over both eyes.

Cleared to continue by New York State Athletic Commission physician Dr. Harry Kleiman, Quarry came for the fifth with his hands held high to protect his face. When Frazier drilled him with yet another unanswered right hand to the head, Quarry held up his gloves in apparent surrender, as blood began pouring harder now from the cut over his left eye. Frazier pointed to Quarry and told Joe Louis, "The man is cut bad. What are you going to do?" But Louis looked at him blankly. Along press row, writers and others called out to the Brown Bomber, "Stop it, Joe! Stop the fight!" Thinking that the action would be halted, Frazier turned back to his corner, but Louis ignored the cries to end the fight and signaled the two fighters to continue. Upon hearing Futch yell, "Go back in there!" Frazier caught Quarry with two left hooks that caused his body to shudder. Only then did Louis step in and stop it, at 1:37 of the round. In his column the following day in the *Los Angeles Times*, Jim Murray waxed grimly, "It was a fight so brutal and atavistic [that] this dispatch should be filed in cave drawings."

Given up for a ghost of his old self only days before by the press, Frazier was full of new energy and resolve in his evisceration of Quarry. Clancy claimed he had never seen Frazier look better and predicted he would beat the winner of the Ali-Foreman bout scheduled for Kinshasa on October 30. "O ye of little faith," Futch told reporters with a big grin. As Frazier had said of Bugner the year before, he had no desire to injure Quarry more than he had. Joe explained, "When it looked like the skin was ripping off his eye, I wanted it stopped. I have a warm feeling for Jerry." On his way out of the Garden, Frazier stopped in and said good-bye to Quarry, who had required a total of eighteen stitches—fifteen over the left eye, three over the right. "Stop by if you ever happen to be in the neighborhood," Frazier told him. With an ice bag affixed to the purple bruises that covered his face, Quarry told him, "You are a helluva man." Dejectedly, he told the press that he had "no strength," fueling speculation that he had left it between the sheets. Or had it come down to the inescapable fact that the "new" Jerry Quarry was a shot fighter before he even stepped into the ring? Of what remained of his career, now scattered in shards by his feet, he added poignantly: "I had this long, elusive dream. I'm not too sure it can be made now."

Even if Quarry had come in somewhat weaker than advertised, Frazier had a new lease on his career. Given his poor performance against Foreman in Jamaica, Frazier had hoped to get another crack at him and win back his championship. But Ali upset Big George in "The Rumble in the Jungle" in an eighth-round knockout, and a rubber match between the two rivals was now inevitable. To get his work in, Frazier hooked up with Jimmy Ellis in Melbourne before an announced crowd of only fifteen thousand at St. Kilda Junction Oval; press reports placed the attendance at less than half of that. Five years removed from the drubbing Frazier had handed him at the Garden, Ellis, thirty-five, had won only one of his previous six bouts. Notwithstanding, he gave Joe trouble in the early rounds before Frazier stopped him in the ninth on a technical knockout in an altogether lackluster showing. Three months later, Ali was in Kuala Lumpur and had just beaten Joe Bugner when he announced what would be one of his many "retirements," only to reconsider when he saw a tape of the Ellis fight; he laughed and said he was sure Joe was finished. But Frazier was far from that. He told reporters: "I want Clay like a hog wants slop."

Weeks before Ali left for the Philippines in September 1975, the writer Nik Cohn dropped in to see him at his training facility in Deer Lake, Pennsylvania. On assignment for *New York* magazine, Cohn happened upon a scene that underscored the genuine hostility that engulfed the impending encounter with Frazier. As a hundred or so spectators looked on from folding chairs that had been set up in rows surrounding the ring apron, Ali toweled off at the end of his workout and turned to the crowd, which he engaged with what had become his now-tired boast of superiority. "I am Ali," he announced. "I outshine the sun." Sensing a certain torpor pervading the room, Ali dug deeper into his repertoire and produced a word that caused the largely white audience to stir.

Gorilla.

Pick up Cohn:

"'Joe Frazier,' said a salesman in the front row, plump and pink.

"'The ape-man,' said his wife, faded sexy, in a halter top."

Cohn continued: "'Gorilla,' said Ali one more time, and he began to disclaim. Mouth angry, eyes blank, he stared above the massed heads, focusing on infinity, and he screamed, he preached, he ranted. He said that Joe Frazier was so ugly, he should donate his face to the Wildlife Fund. So ugly that mirrors paid him not to look in them. Ugly, ugly, ugly. How could we have a gorilla for a champion? What would the people in Manila think, the people all over the world? That all Americans looked like that? That all black brothers were animals? Ignorant, stupid, ugly? If Frazier stood as our champion, what would other nations think of us?

"'Degenerates,' said the salesman.

"'Right on,' said Ali. 'Freaks.'"

Cohn concluded: "The crowd loved every moment. Jammed tight in their rows, they rocked and stamped their feet, roared at each new sally. . . ."

Foreign governments were now eager to get involved with Ali. With the emergence of promoter Don King, the former Cleveland numbers kingpin who had served a prison sentence for stomping a man to death, the seat of promotional power no longer resided at Madison Square Garden. Sporting a Bride of Frankenstein hairstyle, King had solidified an inside position with Ali and his manager, Herbert Muhammad, by producing a five-million-dollar guarantee for the Foreman bout in Zaire, where King and his co-promoters linked up with corrupt President Mobutu Sese Seko.

While the Garden had been in the early running for Ali-Frazier III, King joined forces with rival promoter Bob Arum to bring it to the Republic of the Philippines, where President Ferdinand E. Marcos had declared martial law three years before. By staging an international sporting event—the "The Saga of Our Lifetime," according to King—Marcos hoped to offset the image of the Philippines as a third-world country steeped in crime and soul-crushing poverty. For Ali, the champion, it would be another eye-popping payday: $4.5 million. Frazier, the challenger, would receive two million dollars (plus one hundred thousand dollars in training expenses). At the July 17 press conference held in Manhattan at the Rainbow Room, Ali produced a rubber toy gorilla he had found in a novelty shop and began batting it in the head, as the roomful of guests erupted in glee. Having promised himself (and Futch) that he would not blow his cool again, Frazier looked on impassively as Ali announced, "It's gonna be a chillah and a thrillah and a killah when I get this gorilla in Manila."

Eighteen days before the fight, Frazier embarked from Philadelphia on the twenty-six-hour flight to Manila, with stopovers in Los Angeles and Honolulu. Even before he stepped on the plane, he had trained more than a hundred rounds, punishing the usual lineup of sparring partners to a fare-thee-well. As he worked out, he was serenaded on the loudspeaker by a song he had written, "First Round Knockout." ("He was down, he was down. Lord knows he was down.") "Man," Frazier said, "I love that song." But Frazier said he had no intention of taking out Ali quickly if he could help it, not in light of the hard feelings that had bubbled up again. "If he goes down early, I plan to run over and pick him up," said Frazier, who had been hearing reports that Ali had been amusing crowds in Deer Lake by taunting a large stuffed gorilla. "I just want to beat on him for fifteen rounds." By the looks of him, Frazier appeared in tip-top condition when he boarded the plane, yet Denise Menz remembered he had a scare one morning with his back. Upon waking up in his room at the Philadelphia Marriott, he could not move enough to get out of bed. "I helped him into a hot tub. He could barely walk," said Menz, who summoned orthopedic specialist Dr. Joseph Fabiani to administer a cortisone shot. Concerned that his back would flare up again, Frazier asked Fabiani, "Go with me to Manila, would you, doc?"

But Frazier was plagued by a potentially worse problem as he flew across the Pacific: his eyes. Although his deteriorating vision had been rumored in the press, Frazier had always said that his eyes were fine and the subject was more or less dropped. But as Yanoff had observed, the vision in his left eye had become particularly poor due to the repeated trauma it had received. Menz would remember that the cataracts on the eye were visible. "There were bumps on the surface of the eye, one big one and three small ones," said Menz. "You could see them. It was as if someone had dabbed Vaseline on his eye." Menz said that he was unable to read the small print in a newspaper and that she had to label his eight-track tapes in big block letters. Even so, he did see well enough to somehow obtain a license to drive a car, and he did somehow pass his physical to box. Beyond Yanoff and herself, Menz said only a handful of people in the inner circle knew the extent of the problem, presumably including Futch. Joe himself downplayed the potential peril, telling Menz: "I would rather be rich and blind than poor and blind."

Welcomed to Manila International Airport by a Filipino security force with automatic weapons and sidearms, Frazier told reporters that he was as "sharp as a razor." Typhoon season was under way upon his arrival that September 13, yet an even bigger wind swept into the Philippines two days later as Ali showed up on the red-eye from Honolulu with his entourage in tow. "The great annihilation of Joe Frazier will take place here!" Ali announced from the tarmac, where a member of his receiving committee draped a garland of flowers around his neck. Also on hand to greet him was

Marvis Frazier, then fifteen, who presented Ali with a "Smokin' Joe" T-shirt. Playfully, Ali clenched his hand into a ball and waved it in front of Marvis's handsome young face, telling him: "I'm so anxious to get my fists on your father's nose." Until then, he would have to content himself with his daily barbs at Frazier, of which he seemed to have an inexhaustible supply. Although words never failed him, he occasionally even employed visual accompaniment, flattening his nose with his index finger in a crude characterization of Frazier as a cauliflowered pug. A wire photo of Ali in this pose appeared in papers across the world.

Crowds on the order of three thousand or more Filipinos poured into the Folk Art Center for the workouts. By now, Frazier was going easier on his sparring partners. According to Benton, they were "working on the little things, polishing," with particular emphasis on shoring up his defensive tactics. "Before, his only defense was his offense," said Benton. Otherwise, Benton assured the press that Frazier would pour on the same pressure that had been his hallmark, noting: "The meanness is still there." Publicly, Ali claimed Joe was no longer the same "little, slick-haired" dynamo who had beaten him four years before. Although Frazier was two years younger at age thirty-one, Ali claimed that Joe was actually five years older than him physically due to the many blows he had absorbed through the years. Ali said, "When I start hitting him in the head, it will be the one million, eight hundred thousandth time he has been punched in the head." Privately, Ali knew that he would be in for a long evening and that he could not prepare himself for Frazier in the same cursory way he had for some of his lesser opponents. Observed Gene Kilroy, his business manager: "Ali trained his ass off for Joe Frazier."

From the presidential suite at the Hyatt House, Frazier could look out on the big ships anchored in Manila Bay, beyond which loomed the World War II battlegrounds of Corregidor Island and Bataan. Nine days before the bout, Ali showed up below his balcony and shattered the peace of a late Sunday afternoon by calling out to Frazier: "Gorilla, I want you out of Manila by sundown!" According to Eddie Futch, Joe placed a foot up over the railing of the balcony, as if he intended to hop over it and go get Ali. "[Joe] was just kidding around, but we were nine floors up!" Futch said. Down below, as Ali tried to wrestle a pistol from one of the security people with him, another unloaded his gun and gave it to him. In a piece of silliness that could have easily gone wrong, Ali pointed the weapon at Frazier and began clicking the trigger. Only later did Frazier find himself growing annoyed, telling *Philadelphia Daily News* columnist Tom Cushman: "The more I thought about it the dumber it seemed." At a press conference at the Folk Art Center, Ali produced a toy gun that he said he had purchased on the street from a small boy for fifty pesos ($6.50). Frazier said later, "That was no cap pistol he had with him the other day."

While the gun episode was purely a comic play by Ali for headlines, there was nothing contrived about the quarrel he would have a few days later with his spouse Khalilah. Ali and Frazier had gone with a group of reporters to the Malacañang Palace for a visit with Marcos, who eyed the stunning model Ali had at his arm and complimented him on his "beautiful wife." The woman was Veronica Porche, who had been with Ali in Zaire and Kuala Lumpur and had been the cause of continuing friction between Ali and his actual wife. "All of us cringed when Marcos said that, but no one corrected him," said Patti Dreifuss, one of the publicists for the promoters. Upon learning what happened, Khalilah flew to the Philippines. She found Ali in his suite at the Manila Hilton, where she told him coldly, "We have a lot of things to talk about." Talk they did. For an hour, Khalilah and Ali engaged in a heated row, with Khalilah underscoring her ire by hurling lamps at her roving spouse. In the hallway outside, *New York Times* reporter Dave Anderson had dropped by for an unrelated appointment with Ali, only to spot Khalilah leaving ahead of a bellman with six suitcases in tow. "Nobody wants me here," Khalilah told Anderson, as she stepped on the elevator. Twelve hours after she had arrived, she boarded the exact same plane back to the United States.

Given how Ali had been taunting him, Frazier could not help but be amused. Reporters who saw him in the aftermath said he never looked so relaxed. In fact, he was perhaps too relaxed, given that his own dalliances would have left him similarly exposed. Of Ali, he said: "I guess he must be some superman like he says. I mean, I have enough trouble with one woman." Later, he would playfully introduce Florence as his "wife" *and* his "girlfriend," and allow that she brooked no foolishness from him. When his hometown *Daily News* arrived at the newsstands the following day, it carried a short article on the shipshape status of his marriage under the headline, JOE FRAZIER: 1-WOMAN MAN. Hearing Joe go on, Menz gasped: "What are you, crazy? You are no better than he is. Ali and his people know about me. What if he were to say, 'The Gorilla has some nerve. Look at him over there with whitey.'" Joe shrugged. Although Ali would not discuss the subject of his own infidelity in any depth, shooing reporters away from it by saying his personal life was none of their concern, he hinted that his blowup with Khalilah had been a stunt. "We wanted to fill the newspapers and I imagine we got that done," Ali said. "They told me the [closed-circuit] theaters was dying, so we had to do something." No one was buying any of it.

Along with the humiliation he had heaped upon Khalilah and the awkward appearance with Veronica at the presidential palace, which a Roman Catholic priest later called a "national insult" in a column for the *Philippines Daily Express*, Ali continued to savage Frazier as apelike in appearance and intellect. When asked why he did it, Ali explained it was a form of "mental warfare" to incite Frazier to lose his cool in the ring; and by the way, Ali pointed out, why had Frazier continued to call him by his slave name, "Clay"? And yet even his devotees in the press would agree that the act had become unseemly. By referring to Frazier as a "gorilla," he was casting him as subhuman in the same way that southern plantation owners had looked upon their slaves. Ironically, it was the same type of vile language that Cushman had encountered when he stopped by that tavern on Parris Island and shared beers with a group of marines who called Ali a "nigger." Outwardly, Frazier had more or less held his peace since the studio scuffle, yet he seethed inside. Kilroy said, "Joe was like, you know: 'That dirty motherfucker called me what?'" As he and Ali were leaving the Malacañang Palace, Frazier peered over the top of his sunglasses and told him, "I'm gonna whip your half-breed ass."

The sun was just beginning to come up over Manila Bay on October 1, 1975, when Joe and his handlers stepped from the air-conditioned sanctuary of the Hyatt House for the short walk to his waiting car. Even at that hour the heat and the humidity were nearly unbearable. It reminded him of how it had been when he was a boy in Beaufort, where the asphalt would become so hot at the height of summer that he only drove his car in the cool of the evening to preserve his distressed tires, which he called "may pops." (Why may pops? "Because," he explained, "they may pop at any time.") But Manila was hotter than he had ever remembered it being in South Carolina, and the Philippine Coliseum in suburban Quezon City was hotter still—114 degrees in the ring—due to an inadequate air-conditioning system, a tin roof, and the close proximity of television lights. Along with Marcos and his wife, Imelda, twenty-eight thousand spectators and more than one thousand foreign journalists squeezed into the arena. In sixty-eight countries across the globe, including 380 closed-circuit locations in the United States and Canada, millions more settled into their seats for the closing act of Ali-Frazier, the Thrilla in Manila: according to *Newark Star-Ledger* columnist Jerry Izenberg, fifteen rounds for "the championship not just of the world but of each other."

In the center of the ring sat a gleaming trophy. When the ring announcer declared it would go to the winner and that it had been donated by Marcos, Ali sprinted from his corner, picked it up, and carried it off in the manner of a burglar. The crowd roared in approval at the introduction of Frazier yet greeted Ali with boos. Was it because Ali was Muslim and the Filipinos were a Christian people? Or had his antics and his cruelty toward Frazier become unsettling to them? In any case, Ali turned to the crowd with an expression of shock and feigned tears, rubbing them away with a gloved hand, before shedding his affected dismay and lighting up the building with a sunny smile that converted the jeers to cheers. The referee was Carlos Padilla, a Filipino who had been chosen from a field of three finalists that had included Philadelphian Zack Clayton, who refereed the Ali-Foreman bout in Zaire but whom Futch had opposed due to his close association with King. As Padilla reviewed the rules with Ali and Frazier in ring center, Ali blabbed: "You don't have it anymore, Joe. You don't have it. I'm going to put you away." Frazier grinned and replied, "We'll see."

Ali had told his fans beforehand to take their seats early "because Joe Frazier may sit down before you do!" At the bell for round 1, he came out of his corner intent to capitalize on the fact that Frazier was a slow starter, as he had proved during his career again and again. Nine and a half pounds heavier, at 224½, than he had been in their 1971 bout—Frazier was ten pounds heavier, at 215½—Ali was no longer up on his toes as he had been years before. Moving counterclockwise as Frazier pressed forward, Ali set himself and unloaded heavy right hands. From his corner, assistant trainer Bundini Brown cried, "He won't call you Clay no more!" Near the end of the round, Ali lured Frazier into a trap on the ropes, pivoted to his left, and popped Joe in the jaw with two left hooks. Though the first was a pawing blow that packed only scant power, the second, though it traveled no more than six inches, caught Frazier as he was lunging forward and sent him hurtling into the ropes. Ali followed with a straight right hand to the chin and left uppercut, but Frazier recovered enough to extricate himself from the ropes. In an arms-crossed defense to protect his chin, Frazier charged ahead as Ali battered him with whipping blows.

Off to a commanding start, Ali would more or less have his way through the early going. Extending a gloved hand to hold the charging Frazier at bay—the same "yardsticking" tactic he had used against Liston to buy time in the fifth round of their first bout—Ali snapped off a hard right cross to the chin that caused Frazier to dip toward the canvas. The crowd gasped. But Frazier recovered and plowed ahead, working the body as Ali clutched Joe by the back of the head. Unlike the referee in their second bout, Padilla would have none of it; he stopped the action and gave Ali a warning. Ali opened the fourth round with more "yardsticking" and then unleashed four left hooks, scoring heavily with three of them. He then backed into the ropes, covered up, and summoned Joe to come in and get him, just as he had done against Foreman, in a strategy he would come to call the Rope-a-Dope. Foreman had punched himself to exhaustion in Zaire, but Frazier had promised that he would "embalm" Ali on the ropes if he opted to challenge him there. With Frazier firing away with both hands, Ali stepped to the right and turned Frazier, who now had his back to the ropes as Ali strafed him with right hands. But Frazier battled back, belting Ali with a solid left hook to the head and forcing him into the center of the ring. Late in the fourth round, as Ali used his superior hand speed to beat Frazier to the punch, announcer Don Dunphy observed: "This has been a one-sided fight so far."

On his stool at the end of the fourth, Ali was breathing hard and leaning back on the ropes behind him for support. But prior to the bell, he stood up and pumped both gloves in the air to lead the crowd in a chant of "ALI! ALI! ALI!" For the better part of the fifth round, Ali planted himself on the ropes, as Frazier snapped his head sideways with a short right hand and then caught him with another. Ali asked, "Where did you get that right hand?" Frazier replied, "From George Benton." Ali covered up. Frazier pounded him on deltoids and biceps in order to weaken his jab, which had been another tactic shown to him by Benton. When Ali lowered his gloves, Joe stepped in and pummeled him underneath his heart and to his ribs. From his corner, Angelo Dundee shouted at Ali: "Get off the goddamn ropes!" As Ali opened himself up in the process of throwing a right uppercut, Frazier hammered him with a left hook to the jaw reminiscent of the shot with which he had floored Ali in the fifteenth round of their first meeting. That had also come on the heels of a right uppercut from Ali. Given the peril it invited, Ali had rarely thrown it in their second bout but found himself forced to employ it again as Frazier poured on intense pressure. By the end of the fifth round, the crowd was now chanting: "FRAZIER! FRAZIER! FRAZIER!"

Chroniclers of the Thrilla in Manila would later write that it was a drama in three acts. Ahead on two of the scorecards 3–0–1 and on the third at 3–1, Ali had prevailed convincingly in act 1. But act 2 would belong to Frazier. He captured the fifth round on two of the three scorecards and won the sixth unanimously as he pressed the pedal to the floor. He walloped Ali with a left hook to the head, followed him into the ropes, and then walloped him again with another one. Later in the round, as Ali effected a counterclockwise retreat, Frazier shook him again with a long right hand to the jaw. When Ali came out for the seventh round, he told Joe: "Old Joe Frazier, why, I thought you were washed up." Frazier replied, "Somebody told you all wrong, pretty boy." Up on his toes now, Ali peppered Frazier with long left hands. Even as Frazier remained the aggressor, Ali controlled the action, breaking it with clinches and resuming it as he pleased. Between the seventh and the eighth rounds, analyst Ken Norton observed that Frazier was "in better condition" than Ali, who was breathing hard on his stool in the corner. Round 8 saw Ali plant his feet in the center of the ring and go for the knockout, but Frazier covered up and drove Ali to the ropes, where Joe pummeled him to the body and head. Frazier again drove Ali to the ropes in the ninth and tenth, grunting as he dug in with body shots. By the end of the tenth round, two of three scorecards were even (4–4–2), with Ali ahead on the third (5–3–2).

My father wrote in *Sports Illustrated:* "Ali sat on his stool like a man ready to be staked out in the sun. His head was bowed, and when he raised it, his eyes rolled from the agony of exhaustion. 'Force yourself, Champ!' his corner cried. 'Go down to the well once more!' begged Bundini, tears streaming down his face. 'The world needs ya, Champ!'"

Act 3 began with Ali on his toes again. Frazier backed him into his corner in the eleventh round, where he banged him to the head and body, yet Ali spun back into the center of the ring. Somehow, he found another gear, while Frazier began to flag, both of his eyes beginning to swell. From his seat at a closed-circuit venue in Philadelphia, his ophthalmologist began to grow worried. "At the time, I was one of the very few people in the world who knew how poor his vision was in his left eye," Yanoff said later. "I was horrified when, toward the tenth round, his eyelids started to swell over his one good eye, his right eye. In the last three rounds, the lid was swollen shut. He could see only out of his mostly blind left eye." With Frazier unable to see, he labored to get inside, where he could track Ali by the sound of his breathing. Through the twelfth round, Ali scored heavily as Joe was now bleeding from the mouth. Midway through the thirteenth, Ali pivoted to his right and began dancing away from Frazier, counterclockwise. As Joe followed, Ali battered him with a left-right combination that sent his mouthpiece sailing into the sixth row of the ringside seats. A chopping right appeared as if it would topple Joe in ring center, but Ali was too exhausted to press his advantage. Ali landed punches at will as Joe plodded forward, feeling his way.

From his vantage point in the corner during the fourteenth round, Eddie Futch could see what Yanoff had spotted from more than eight thousand miles away: the end was near. Right hands caromed off Frazier, sending sweat flying into the ring lights. By now the blows of both men were landing even harder, as their gloves had become waterlogged to the consistency of wet cement. With his field of vision compromised, Joe became a stationary target. "Frazier is hurt!" Dunphy alerted. "Frazier is badly hurt." Exhausted himself, his arms now heavy, Ali could not bring Frazier down. Joe would not capitulate, not now or ever, not if he had a single breath remaining in his heaving lungs. Futch knew that the outcome could be unforgiving. He had seen seven fighters die in the ring, including that one in Detroit in 1949: Talmadge Bussey, whose brothers had squabbled over whether to send him back out to meet his fate in the ninth. One had wanted to throw in the towel, the other had wanted to gamble and give him another round. For Futch, in this moment that called upon him to display nothing short of moral courage, there was never a question which brother he would be. Guided back to his corner by Padilla at the end of the fourteenth round, Frazier sat on his stool, bracing himself on the ring rope behind him. He was breathing hard as Futch leaned down to speak to him.

"Joe," he said, "I'm going to stop it."

"No, no, Eddie, ya can't do that to me," Joe replied. He began to rise.

"You couldn't see in the last two rounds. What makes ya think ya gonna see in the fifteenth?"

"I want him, boss."

"Sit down, son," Futch said. "It's all over. No one will ever forget what you did here today."

Ali stood up from his stool and raised a weary arm in victory.

Surrounded by Dundee, Pacheco, Brown, and the others, he collapsed in exhaustion to the canvas. Someone helped him back onto his stool. Others began fanning him with towels. Ali sat with his head down, his elbows on his knees. With his gloves now off, Frazier walked over and told him solemnly: "Good fight." Dunphy leaned in and asked if there had been a point where he had any doubt that he would win. "Round 10," said Ali, still breathing hard. "I was surprised Joe had so much stamina." President Marcos found his way to Ali, congratulated him, and presented him with the trophy that was now legitimately his and his alone.

When Bob Goodman came to get Joe for the press conference, he found him sobbing on a bench in his dressing room. Futch had his arms around his shoulder, consoling him. "Guys, you have nothing to be ashamed of," the publicist told them. "It was one of the greatest fights in history. Come on, the press is downstairs." His eyes embedded in swollen flesh, Joe looked up and replied, "Okay. Get my shoes." Quickly, Goodman stepped away and hurried across the hall to convey this piece of information to Ali, who had previously told him that he was in no shape to attend. Goodman would say years later, "He was stretched out on a couch and could barely move. I had never seen him so exhausted."

Goodman now told Ali, "Muhammad! Muhammad! Frazier is going. And he lost!"

Ali perked up and said, "Frazier is going?"

Gingerly, Ali lifted himself up and told Goodman, "Get me my comb." Goodman would remember, "Ali always had to look good."

Wearing a pair of sunglasses perched atop his puffy face, Frazier was gracious in his praise of Ali. "My man fought a great fight," he said. When he was asked if he was upset that Futch stopped it, he said candidly that he was "not very pleased about it," quickly adding: "But as you know, Ed is the boss." Another reporter asked:

"How beat up were you?" Frazier replied, "Not beat up. . . . My eye closed on me." *New York Times* reporter Dave Anderson said years later, "He sat there and took his questions and was very cognizant of everything. Then everybody began to wonder, 'Where is Muhammad?' . . . It was a good forty-five minutes before Ali came out. And when he did, he just kind of wobbled out and sat down. And somebody asked—I think it was the first question: 'What was it like in there?'"

"Gentlemen, what you saw tonight was next to death," he said. "I always knew Joe Frazier was great, but he was even greater than I thought he was. I could not take those punches he took tonight."

As Ali fielded questions, it occurred to Anderson: If you had just come off the street, looked at the two men and had to guess which won the fight, "you would have thought, 'Hey, Frazier won it!'" Had you arrived at that conclusion, it would only be because while Frazier looked as if he had been worked over in an alley on the wrong side of Manila, Ali called to mind a colorful kite that had gotten caught on a high-tension wire, its tangled pieces flapping in the wind. Quietly, Ali said: "My arms hurt. My hips hurt. My eyes hurt. My brain hurts. Everything hurts." He said the late rounds were so hard "you want to throw up." Asked what was next, he said, "There is a great possibility that you saw the last of Ali. . . . Let Norton, Foreman, Frazier, or whoever fight for the title. I want to get out of it." *Newark Star-Ledger* columnist Jerry Izenberg told me, "It was such a brutal, brutal fight that Ali pissed blood for a long time."

Ali attended the celebration held in his honor at the Malaca-ñang Palace. Anderson described the scene: "It was gorgeous. You walked in and there were two staircases going up to the second floor. And in the area between the staircases, there was this huge mother-of-pearl chandelier." While Don King danced with Imelda Marcos to "Tie a Yellow Ribbon Round the Ole Oak Tree," Ali "kind of shuffled into an upstairs room and sat in a chair very stiffly," according to Anderson. "He just sat there with his hands cupped in front of him, as if he were holding a small bird. And you could see he was sore. He could just hardly move. And everybody came up to talk to him and, you know, he could not even shake their hands. He just kind of touched their hands. He looked up at them and smiled, hardly said a word other than 'hello' and 'thank you.'" In his piece for *Sports Illustrated*, my father captured Ali as he was eating from a plate of food that Imelda had earlier filled: "Ali never appeared so vulnerable and fragile, so pitiably unmajestic, so far from the universe he claims as his alone. He could barely hold his fork, and he lifted the food slowly up to his bottom lip, which had been scraped pink. The skin on his face was dull and blotched, drained of that familiar childlike wonder." Every organ in his body had been pulverized.

Upon seeing her father immediately after the fight, young Weatta Frazier and her siblings were alarmed. "We were in tears," she said. "Except for Marvis. You know how he is. He said, 'Pop'll get him next time.'" Frazier told his children, "Nothing to worry about. Daddy is fine." Even with the bruises that covered his face, which he soaked in a bathroom sink full of ice back in his suite, Frazier showed up later at the party held for him on the roof of the Hyatt House full of cheerful exuberance. He jumped onstage, grabbed the microphone, and asked the audience, "What's happening?" Along with the band, he sang an old favorite, "Knock on Wood." Then he posed for photographs and did a spin on the dance floor before calling it an evening. The following day, he held court with reporters as he reclined on his bed. Both of his eyes were swollen to the point that he was unable to see who had asked him a question. Denise Menz would remember, "He told me to call out their names so he knew who he was addressing." In his coverage in *Sports Illustrated*, my father observed: "The scene cannot be forgotten: This good and gallant man lying there, embodying the remains of a will that had carried him so far—now surely too far."

Even so, Frazier would think he had not been permitted to go far enough, that if he had only been allowed to come out for the fifteenth round, the outcome could have been written far differently.

Given the long and complicated history of his warfare with Ali, the ending did not go down easily for Frazier, even if he concealed that in Manila. At one point later, Izenberg said that Benton had told him that when Eddie had advised Joe in the corner that he was stopping the fight, Frazier stated he would never talk to him again if he did. Even more upset was Menz. Of Futch, Menz would say more than once: "If I had had a gun, I would have shot him. Joe had worked so hard and had come so far and now to have it taken away from him, like, with just one round to go? You know what I mean?" Contrary to the widely agreed-upon opinion that Frazier was in big trouble by the end of the fourteenth, she contended that Ali was in no better shape than Joe and that neither had enough power left to inflict any substantial harm. "Ali was dead on his feet. So was Joe. What were they going to do to one another at that point?" Menz said. So noisy behind the scenes in her criticism of Futch, Menz conceded that Joe would become exasperated with her and growl, "Let it go, would ya?" But he himself would not let it go, not in that place deep inside himself where he contemplated alternate realities. Even though he was behind on all three of the scorecards and needed a knockout to win, Frazier could not help but believe Eddie had denied him his chance to pull it out. He was certain Yank Durham would have sent him out for the fifteenth.

Fueling any lingering belief that he had been somehow short-changed was a story that went around that if Frazier had not been preempted by Futch, Ali would have tossed in the towel. A quote by Ali in the 1991 oral history *Muhammad Ali: His Life and Times*, by Thomas Hauser, appeared to give energy to the speculation: "Frazier quit just before I did. I didn't think I could fight anymore." Can we take him at his word? Was he actually thinking of quitting? Or was Ali simply acknowledging the extreme to which both of them had been pushed? Only he himself would ever know, yet there are those who say to this day that they saw evidence of precisely that. Marvis Frazier told me that he and Cloverlay middleweight Willie "The Worm" Monroe were standing in the aisle not far from Ali's corner and were convinced he was on the verge of packing it in. "We were waving across the ring at Mr. Futch not to stop it, but we could not get his attention," said Marvis. Gordon Peterson, a publicist assigned by King to the Frazier camp, said that "when Ali came back to the corner at the end of the fourteenth round, he could barely walk. Of course, we will never know, but I would not have been surprised had he not come out." He was not the only one.

And yet others would adamantly argue the point. "Total crap," said Gene Kilroy, who was in the corner. While Angelo Dundee was still alive, he always discounted it as fiction. Given how he had shoved Clay out of the corner with his afflicted eye in the first Liston bout, it seemed highly unlikely that he would not have done the same in Manila when the bell called his fighter out for the fifteenth round. "Never happened," Dundee said. *Philadelphia Daily News* columnist Tom Cushman told me that if there was a juncture where Ali was poised to quit, it occurred at the end of the tenth round, when Ali himself said he was uncertain if his legs would go out from beneath him and he would faint. Cushman observed, "It was the tenth, not the fourteenth, when Angelo more or less pushed him out of the corner." Goodman concurred, saying more than forty years later: "Each round it became harder and harder for him to get off the stool. How in the hell he was able to do it is beyond me. But no, Ali would not have quit under any circumstances: if Joe was still standing to this day, Ali would still be standing there."

Only seasoned observers of the sport would truly appreciate the act of bravery that Futch had performed on behalf of Joe. Given the big stakes that were on the table, it would have been very easy for him to overlook what his eyes were telling him and give Joe that final round. By stopping it when he did, Futch very easily may have saved Joe from an untimely and tragic death. "I think Joe would have won it if his eye had not closed up on him," said Yanoff. "But as he could not see out of it, I think Futch probably saved his life." For Larry Hazzard, the New Jersey Athletic Control Board commissioner who years before had boxed as an amateur with Joe, Futch "endeared himself forever to anybody who had any feelings for Joe Frazier" by the preventative action he took. Having worked as a referee for years, he observed: "I know a lot of guys who would have sent Joe back out there. Joe himself would have done it. See, fighters have a gladiatorial spirit. Real fighters, 'quit' is not in their vocabulary. He *had* to go back out there. But the punishment those two were putting on each other? In my opinion? Someone was going to die." Although Hazzard was not in Manila, he watched the event on closed circuit. "I was screaming for somebody to do something," said Hazzard, who has had losing fighters thank him years later for halting a bout while he was a referee. "Well, Eddie Futch did. And I tip my hat to him."

Never once would Futch second-guess himself. Whenever word got back to him that Joe was displeased with him, he reminded himself that he knew what he knew: Had he allowed Joe to come out for the fifteenth, there was a chance that with one more shot to the head, Florence would be seeing to funeral arrangements. Although he had no way of knowing then if Joe was ahead or behind, it would not have factored into his decision either way, the same as it had not been a consideration when he was chief second for welterweight Hedgemon Lewis in his 1969 bout against Ernie "Indian Red" Lopez. "There were only thirty seconds to go in the last round," Futch told Anderson for his *New York Times* column, "and I was sure Hedge was ahead. But I had to stop it." In his conversations with Frazier, Futch told him, "Joe, you are still a young man. You have a wife and beautiful children who love you dearly. You have your whole life still ahead of you. Did you want to throw all that away for just a fight?" On a certain level, Frazier could appreciate the place in the heart from which these words had come, yet on a deeper level, where ego collided with good sense, he only knew what *he* knew: It had not been "just a fight." It had never been that. Yank would have understood that. He would have pushed Joe out of the corner for the fifteenth and told him, "Go get him, cocksucker!" Even as Joe clung to this belief, Futch always knew better. He had given Yank his word and he had kept it: He had taken care of his boy.

CHAPTER TEN

BOOGIE, BOOGIE, BOOGIE

Joe on stage, 1971.
Getty Images

Eddie Futch once told Yank Durham they would have to keep a close eye on Joe as he grew older. "Guys like Joe, when they go out, they go out quickly," Futch had said. Given the walk-in style Frazier had adopted, he had absorbed thousands of head shots through the years, which had accumulated in ways he could not begin to calculate. With an eye toward sparing him any more wear and tear than necessary, Durham arranged for him to get into his own training facility to steer him away from the brain-rattling warfare that played itself out in the call-to-manhood battles at other Philadelphia gyms. To further protect him, Durham and later Futch saw to it that he had no more than two bouts per year beginning in 1969, even though Joe would become edgy and hector them for more. "Get me a fight," Joe would plead. "If not a big one, then get me a small one." Had Durham lived beyond 1973, it seems possible that he could have coaxed Frazier into the safe harbor of retirement with perhaps more persuasiveness than Futch. Chances are that when Frazier came back from Manila, there would have been no talk of stepping into the ring again with George Foreman, who had beaten Joe three years before as if he were a three-egg omelet. Sagely, future heavyweight champion Larry Holmes told me years later: "You can't keep taking those punches upside your fucking head."

Even as George Benton said he would be happy if he never earned another dime off him, Frazier had no intention of packing it in. Beating Foreman would enable him to clean up some old business he had with Big George and perhaps pave the way for a fourth shot at Ali, who quickly withdrew his Manila "retirement" announcement and went on to inconsequential wins in 1976, over Jean-Pierre Coopman, Jimmy Young, and Richard Dunn. Though Frazier would not agree with Futch that he had come in "half prepared" when he faced Foreman in Jamaica, he got upset whenever he brooded upon the shellacking he'd received. Frazier understood why Benton and others would have preferred he retire, but as he told a United Press International reporter: "What am I going to do, sit around and watch old films and wait for a heart attack?" Only *he* would know when he'd had enough. As he told Dave Anderson in the *New York Times*, "In the morning, nobody has to touch me with an electric pole to get me up to do roadwork." Still, by the end of January 1976, it was not yet clear if Futch would sign off on a Foreman rematch. "I want to check him out against a fair sparring partner," said Futch. "I want to check his legs, his reflexes, see how he compares with the Joe Frazier I've known for the last ten years." Futch gave his approval, yet knew the end was near.

Eleven years had passed since Joe turned pro. Faces had come and gone: Joe Hand Sr. had stepped away from Cloverlay prior to the first Foreman bout due to some political infighting in the organization and launched what would become a highly lucrative closed-circuit enterprise; Lester Pelemon grew weary of the "party life" that surrounded Joe and headed back to Syracuse for a job at Kodak. Cloverlay itself ended its agreement with Frazier on the eve of the Thrilla in Manila, which enabled him to take home the full portion of his purse without splitting off 25 percent to its nine-hundred-plus shareholders. For the forty charter investors who purchased eighty shares at $250 apiece and saw their shares split, a venture that began as an act of civic-mindedness became a financial windfall upon liquidation.

Even with the percentages that he doled out to Cloverlay, Frazier had done far better financially than he could ever have dreamed. "The love" had poured in. Of the more than ten million dollars he earned in purses, he ended up with net earnings, less expenses and taxes, of $3.1 million (just under $25 million in 2018 dollars). He owned a parcel of land in Bucks County, Pennsylvania, for which he had been offered $2 million, the $350,000 plantation in Beaufort County, South Carolina, and a $150,000 home outside of Philadelphia in Whitemarsh Township. Additionally, he had a pension fund at Provident Bank that contained $500,000 and a portfolio of stocks and bonds valued at $300,000. According to attorney Bruce Wright, there were "another couple of hundred thousand in cars, clothes, and miscellaneous properties." Cloverlay also sold Frazier the gym on North Broad Street for $75,000. At the still-young age of thirty-three, Frazier appeared to be set for life. Unlike Joe Louis and an array of other champions through the years, Frazier did not *have* to fight. But he and Foreman were offered $1 million each for their rematch, which was no small piece of "love" given what he conceded were his "expensive tastes."

Worsening eye problems remained an obstacle to moving forward with his career. Upon arriving back in Philadelphia from Manila, Frazier scheduled an immediate appointment with Dr. Yanoff for an examination. Records indicate the vision in his "good" right eye was only 20/50. His left eye was "legally blind." Said Yanoff: "I told him he would have to have the cataract removed from his left eye." For the November 1975 surgery, Yanoff accompanied Joe to New York to see Dr. Charles Kelman, who had developed the small-incision phacoemulsification ("laser") technique to perform the cataract procedure. Concerned that word of his appointment would get out, Joe shaved his head and wore sunglasses as he and Yanoff strolled on Fifth Avenue. Yanoff remembered, "Almost immediately, people started yelling, 'Hey, Joe, what's happening?'" Although the surgery was successful, Kelman did not equip Frazier with the still somewhat experimental lens implant. Consequently, he was forced to wear a contact lens in order to see out of his left eye. Aware that he would be prohibited from boxing if his impairment were discovered, Frazier employed some clever chicanery to beat the eye exam in his prefight physical. According to Yanoff, Frazier "would read the eye chart with his good right eye, memorize it, and pretend to read with his left eye, which was practically blind without a contact lens."

To what extent was Futch aware of that? Futch said years later, "If Joe had been blind, I would have retired him. I would not want any fighter to risk his life because of a physical abnormality." Given that Futch had halted the Manila bout and had saved Joe not just from Ali but from himself, it would be illogical to think he would send him in with one eye against Foreman, who was one of the hardest punchers in history and who had proved it in their first fight. Had Futch been aware of Joe's handicap, there would not have been a second Foreman fight or perhaps not even the Thrilla in Manila. But with his career on the line and at enhanced peril, Joe concealed the scope of his optical impairment from Futch, just as he would hide a variety of private matters from others through the years. According to Denise Menz, Futch had no knowledge of the surgery that Kelman had performed, nor was he aware that Frazier was wearing a contact lens during his sparring sessions. "Other than Florence, no one knew but the two of us as far as I understood it," she said. "He used to carry them around in his shaving kit with his solution. He sparred with them and kept thinking he could keep them in. But even with headgear on, they popped out whenever he was hit and ended up on the floor." Menz remembered that Joe would go through boxes of them. "There were a dozen in each box and they were expensive. They cost something like a hundred dollars for each lens." Frazier acknowledged in his autobiography that he wore a contact lens in his left eye during the Foreman rematch.

Jerry Perenchio joined with Caesars Palace to promote the scheduled twelve-round bout. Madison Square Garden had already been booked for the Democratic National Convention. Perenchio hoped to stage the fight in the refurbished Yankee Stadium, but Yankees owner George Steinbrenner balked in fear that the event would destroy the playing field. Perenchio settled for Nassau Coliseum in Uniondale, New York. Leading up to the June 15 date, advertisements displayed Frazier and Foreman in Roman gladiatorial garb, circa third century B.C., accessorized with laurel wreaths, gauntlets, greaves, and bucklers. In a corny salute to the bicentennial, the two also appeared in television commercials dressed as historical figures from American history, including one in which Foreman was done up as George Washington and Frazier as Betsy Ross. A public sparring session was held eleven days before the bout off Forty-Fourth Street in Shubert Alley, where chorus girls from the show *Very Good Eddie* peered out of windows as Frazier, according to Stan Hochman in the *Philadelphia Daily News,* "did the clumsiest 45 minutes on Broadway since Marlene Dietrich toppled into the orchestra pit." Hochman continued: "If boxing is a jungle, George Foreman is Dr. Livingston. He was due in New York on Tuesday. He was supposed to do the sparring in Shubert Alley. Frazier subbed and got hit with everything but an overtime parking ticket."

Contrary to the avuncular persona he would engage years later as the pitchman for the wildly successful George Foreman Grill, Foreman emerged from his undoing at the hands of Ali in Zaire as a sullen and altogether confused man, his vast potential undermined by self-doubt and suspicion. Held in a promotional headlock by Don King, he embarrassed himself by taking on five opponents in separate three-round bouts in April 1975 at Maple Leaf Gardens in Toronto. The so-called Fainting Five included journeymen Alonzo Johnson, Jerry Judge, Terry Daniels, Charley Polite, and Boone Kirkman. Howard Cosell called it a "carnival" on ABC-TV. Polite had even harsher words, correctly stating that Foreman had "really degraded himself." He added, "An ex-heavyweight champion should act with dignity and not like a clown." The following January, Foreman scored a fifth-round knockout over Ron Lyle, but not before Lyle floored him twice in the fourth.

Seeing Big George fall to his knees and later flat on his face was a revelation to Frazier. Close to two years had passed since Ali had revealed Foreman's shortcomings in Africa, and it was clear to Joe that Foreman still had them. Though dangerous when he connected, he "pushed" his punches instead of snapping them off. Moreover, he wore down over a period of rounds and became vulnerable. Futch and Benton devised a strategy whereby Frazier would stay away from Foreman in the early rounds, not too far but on the outer edge of his range of power, and then go get him. Hochman observed: "It seems as if Frazier is training for two different fights. Caution for four or five rounds and then switch gears, out of reserve and into high as soon as Foreman starts to unravel." Benton tapped a finger on his temple. "The Foreman fight is the kind you win with your fists but you lose up here," he said. "If Foreman gets all of Joe early, we could wind up back in Jamaica." Benton said that he and Futch believed Frazier still had enough ability and heart to win but added: "His body has to give him the real answer. And the only way to ask the question is to climb through them ropes."

Cheers from the 10,341 fans in attendance swelled within Nassau County Veterans Memorial Coliseum as Frazier entered the ring in a yellow robe, accompanied by Futch, Benton, and son Marvis. Once inside the rope, Marvis flipped back the hood of the robe to reveal that Joe had shaved his head, which caused Cosell to gasp at ringside: "The beard is gone! The hair on the top of his head is gone!" At 224½ pounds, Frazier tipped the scales ten pounds heavier than he had been for Foreman in 1973 and 8½ pounds more than he had ever weighed for a fight. Foreman was also 224½ yet towered over Joe during the prefight stare-down at ring center, their eyes locked on one another with cold hostility. Once an appealing young man who had won over the American public by waving the Stars and Stripes at the 1968 Olympics in Mexico City, Foreman was booed when he was announced to the crowd that evening on Long Island. Las Vegas made him a 7–5 favorite.

At the opening bell, Frazier advanced in a crouch, his shaven head bobbing up and down. Circling Foreman, Frazier stayed just outside his punching range, too far away to connect with any punches. Big, strong, lumbering George tracked him, throwing heavy jabs, right uppercuts, left hooks, and straight rights as Frazier covered up and blocked them with his arms. Foreman had pummeled Frazier with right uppercuts in their first fight, but now Joe held his right hand under his chin in order to catch them. Curiously—yet perhaps not so, given that he was wearing a contact lens in his left eye and hoped to keep it from jarring free—Frazier very rarely engaged Foreman. Instead, he stood with his head turned at a forty-five-degree angle, so that his good right eye was in front of Foreman. Through the early rounds, Frazier followed the strategy that Futch and Benton laid out for him by attempting to stay off the ropes and out of the corners. As Foreman piled up points in the first three rounds, the question remained: How far could he go before his stamina gave way?

Some of the sharpest exchanges in the bout happened in the fourth round. Foreman now trapped Frazier on the ropes repeatedly and caught him with left jabs and left hooks. Though Joe ramped up his offense, scoring with exploratory jabs, left hooks, and an occasional right hand, none of his punches appeared to hurt George. Near the end of the round, Foreman unloaded a series of heavy blows, which Frazer followed with two solid left hooks to the head that drew cheers from the partisan crowd. The officials evenly split the round, with one scoring it for Foreman, another for Frazier, and the third calling it even. As Frazier sat on his stool between rounds, he had swelling above both eyes. Meanwhile, Foreman was breathing normally and showed no signs of wear, as he had at the same juncture against Ali in Zaire.

Midway into round 5, Foreman trapped Frazier on the ropes and once again began to unload, leaving Joe clearly shaken by two hooks to the head. Two more hooks and a big right hand to the chin stunned Frazier, whose left eye was now swollen. In his autobiography, Frazier would remember: "George was targeting it. One of his punches knocked the contact lens off the pupil and troubled my vision." A left-right-left combination by Foreman dislodged the mouthpiece from Frazier, who stood his ground and returned fire. Foreman then scored with a potent right uppercut, which straightened up Joe, followed with a left hook to the jaw that caused Frazier to stumble to the canvas on all fours. Frazier rose almost immediately as referee Harold Valan gave him the mandatory eight-count. Foreman looked on impassively from a neutral corner.

As Valan waved the men back together, Foreman moved in again quickly. On wobbly legs, Frazier sent out a couple of wild left hooks, which Foreman avoided. Foreman drove Frazier back toward his corner with head and body shots, once again straightened out of his crouch with a left hook to the head, and followed immediately with a sweeping uppercut to the chin. The blow elevated Frazier off his feet and dumped him in his corner, his back propped against the padded ring post as he sat on the canvas. With his gloves on the top strands of the ropes, Foreman stood over his fallen opponent and stared out at the crowd until Valan escorted him to a neutral corner. Bleeding from a cut over his swollen right eye, Frazier woozily grabbed the bottom strand of both ropes and dragged himself up at the count of seven. As Valan continued the mandatory eight-count, Futch hopped up on the ring apron and hurried along the ropes.

Valan caught his eye and asked, "What about it?"

"That's it," Futch said.

Valan waved his arms over Frazier at 2:26 of the fifth round. Young Marvis Frazier leaped into the ring and rushed to his fallen father. Tears welled in his eyes as he asked him, "Are you okay?" Joe nodded yes, his eyes vague. As Marvis helped him back to his corner, blood splattered onto his cheek and dripped onto his yellow shirt.

As the crowd filed from the arena, George Benton summed up what happened: "Foreman was too much of a mountain . . . too damn big." Foreman himself would say that it had come down to styles, observing: "I could fight Ali a hundred times, he would beat me a hundred times. I could fight Joe a hundred times, I would knock him out a hundred times. And yet Ali and Frazier could fight each other a hundred times and it would be life and death every single time." In his own autopsy of the bout, Futch said, "Joe fought as well as he could until the last thirty seconds of the fight. I told him to stay away for four rounds, and then get in tight when George slowed down. George started getting sloppy in the fourth." But Futch added, to his dismay, that Joe got pinned on the ropes in the fifth round and that was it. In the dressing room immediately afterward, Futch told Frazier in no uncertain terms: "Announce your retirement tonight."

Joe agreed.

Sure, it was "sad to see a gallant guy like Joe go out," Futch conceded. And yet, he added: "All careers come to an end." On the upside, he added that Joe "still has his marbles and is financially secure." At his press conference, Frazier announced that he would now "nail the gloves to the wall and boogie, boogie, boogie." He met the press again the following morning, stopping on the way out of his hotel suite to catch a glimpse of himself in the mirror: his shaved head was covered with lumps, and a patch covered five stitches alongside his right eye. Downstairs by the pool, he assured the gathered writers, "There'll be no more fights." To which he added, "I'll just go back to the gym and watch the young men grow."

———

On the heels of his victory over Ali five years before in the Fight of the Century, between his stop at the White House and his abysmal European tour with the Knockouts, Joe stopped in for an appearance with Dinah Shore on her daytime TV show, *Dinah's Place*. In the greenroom following the program, during which he and Shore whipped up a batch of mashed potatoes and sang a duet of "(Sittin' on) The Dock of the Bay," Frazier sat with reporter Joan Crosby for an interview in which he pondered the hopes he had for his then-eleven-year-old son, Marvis. Unequivocally, he said there was no way he would allow him to become a boxer. Beyond the physical perils that accompanied the career he had followed, Joe observed that it was just too lonely. "A fighter lives like an animal," said Frazier, who spoke of the hardship of being secluded in training camps for weeks on end. Given the choice to do it over again, he would have chosen a profession that did not take him away from his wife and children for as long as boxing had. For someone who found himself increasingly restless at home, it seemed to be an odd statement, doubly so in light of the wealth that boxing had brought him. Even so, if he had an aspiration for his sons, it would have been that they "go some other way."

Quite a fine athlete from an early age, Marvis excelled in junior high school in football, basketball, baseball, and wrestling. But when he began to struggle academically, Joe and Florence transferred him to Wyncote Academy, which did not have an athletics program. Ostensibly, he began showing up at the gym in the fall following the second Foreman bout as a way to stay in shape, confining himself to hitting the bags. Soon, he was asking his father, "How about a fight?" Joe narrowed his eyes and replied, "No, this is no plaything." When he asked a year or so later and his father gave in, he stopped his sparring partner in the second round. Though he assured his disapproving mother that boxing was "just a hobby," it quickly became far more than that to him. It drew him closer to his father, gave them a commonality they would not have had otherwise. "At first, I came in the gym and messed around," Marvis would later tell Stan Hochman. "But I could soon see that it brought us closer together. Now, he was here with me, helping me out. He was spending more individual time with me, away from Mom and my sisters. So much of the time, when I was younger, he was out on the road, making the bread. I know how it sounds, selfish maybe, but I wanted him closer to me. But most of all, I wanted him to be proud of me."

Now, at the end of his career, it had to seem to Joe as if it were beginning again. When he looked at Marvis, it was as if he were looking at a still-flowering version of himself. Even Joe had to say with a chuckle, if Hollywood ever did his life story, they would not have to look for an actor who looked like him to play his part. He pointed to Marvis and said, "They got a guy . . . identical." And yet that was true only to a point: the 185-pound Marvis was taller, at six foot one, and had grown up in an atmosphere of wealth and privilege, far removed from the disadvantaged circumstances that his father had climbed out of in Beaufort. With big brown eyes and a wide smile, he conveyed a soulfulness that seemed to argue against a successful career in the ring, which has always been the purview of young men hardened by need. Amateur heavyweight Jimmy Clark would comment that Marvis was "subsidized just like the Russians. All he has to do is stay in shape and dream." Upon hearing that, Marvis replied that he "kind of resented it." While Marvis conceded that he did not have to work in a slaughterhouse "like Daddy did," he pointed out that he also had responsibilities, noting: "I go to school, just like Jimmy Clark does. I have choir rehearsal every Monday, Bible classes every Wednesday, homework and chores around the house. I cut the grass, take out the trash, feed the dogs; anything my mom and daddy think I ought to do. And I have to train like any other boxer." By then a student at Plymouth-Whitemarsh High School, he passed up playing school sports to remain focussed on boxing. Though he glided up to the door in a Cadillac Seville his pop had given him, he worked just as hard as if he had just stepped off a city bus.

Near the corner of North Broad Street and Glenwood Avenue, the three-story, twenty-thousand-square-foot building that housed the gym dated back to the 1890s, when it was used as a factory that produced window sashes and blinds. Railroad cars clattered by at any given hour, stopping to discharge and board passengers at the North Philadelphia station. In the underpass below the tracks on North Broad Street, hookers would congregate in the shadows and yell across the noisy traffic at Joe when he parked his car on the sidewalk outside the gym door. "Heeeyyy, Smoke! How ya doin', honey!" Some of the top trainers around set up shop at the gym, including Howard and Quenzell McCall, Sam Solomon, Willie Reddish, Bouie Fisher, and Milt Bailey, who in the early years convened high-stakes crap games once the work of the day was finished. Along with them came a plethora of talent over the years, including Larry Holmes, Ken Norton, Leon and Michael Spinks, Jimmy Young, Randall "Tex" Cobb, Matthew Saad Muhammad, Dwight Muhammad Qawi, Bennie Briscoe, Willie "The Worm" Monroe, Eugene "Cyclone" Hart, Bobby "Boogaloo" Watts, Stanley "Kitten" Hayward, Meldrick Taylor, Tyrell Biggs, "Smokin'" Bert Cooper, Tyrone Everett, and Jeff Chandler. With gang activity then rampant in the city, the gym served as a safe haven for the youth of the community who, even if they were not talented enough to be pro fighters, found that the workouts acquainted them with structure and discipline. Joe said of the gym, "I keep it here for the boys. We got so many. I lose count. Not professionals. Some not even good amateurs. Just neighborhood boys. . . . One gym like this does more than a whole squad of cops."

Upon opening its doors in 1970, the Cloverlay Gym was the "nicest, cleanest gym in Philadelphia," according to promoter J Russell Peltz. Cloverlay had invested $160,000 in renovating the building and constructing the gym itself, which would be called "the Cloverlay Hilton" by virtue of its posh appointments. Even before the Thrilla in Manila, Joe began setting himself up for the day he would retire by forming Joe Frazier Incorporated. With the assistance of attorney Bruce Wright and Eddie Futch, he planned to help aspiring young boxers the same way he had been helped by Cloverlay. "I got good treatment from a lot of people, who saw to it I got something from fighting," Joe said. Along with the array of amateur talent he hoped to groom, he arranged to purchase the contract of former Olympian Duane Bobick, of whom Hochman observed: "Eleven brothers. From Bowlus, Minnesota, of all places. Two bars, three stores, 275 people, a blinking yellow light. White. Handsome. White. Young. Strong. Polite. White. Ambitious. White. White." While Frazier would have other pros in his stable early on—including Willie "The Worm" Monroe, who had beaten a young Marvin Hagler at the Spectrum in Philadelphia in 1976 but came to feel overlooked by Frazier in years to come—

none of them possessed the earning potential of Bobick, the latest incarnation of the Great White Hope. Whether he had the talent to capitalize on it was quite another question.

Whatever the redheaded young heavyweight lacked in ability, he charmed the press with an abundance of personality. He told Douglas S. Looney of *Sports Illustrated* that he began boxing "when the doctor slapped me on my rear and I hit him with a left hook." Growing up in poverty in rural Bowlus, he came from the same hardscrabble environment that had spawned Jerry Quarry. He helped his father dig graves and found himself in more than an occasional donnybrook with his brothers. He became a champion boxer in the navy and beat the premier Cuban boxer Teófilo Stevenson at the 1971 Pan American Games, only to lose to him in the quarterfinals a year later in the Olympic Games in Munich. He turned pro in 1973 under Denver cable-television pioneer Bill Daniels, a former World War II and Korean War fighter pilot who signed Bobick for a bonus of twenty-five thousand dollars and 50 percent of his gross purses less expenses. For both owner and fighter, it was far from a remunerative partnership. While Bobick ran off twenty-five consecutive victories under Daniels, he did so for loose change against opponents that *St. Paul Pioneer Press* columnist Dan Riley called "old ladies, roundheeled has-beens and clowns moonlighting in the Shrine Circus." Bobick opted out of his contract for a fee of between $107,500 and $150,000. Daniels said he was sorry to see him go.

On the recommendation of Wright and with the approval of Futch, Frazier signed Bobick to a contract that would pay him 47 percent of his gross purses in addition to a pension benefit; Joe would pick up the tab for boxing-related expenses. To assess his potential before inking the agreement, Frazier banged Duane around in some sparring sessions. Although he liked what he saw well enough, he did not fall in love with it: Bobick had a passable straight right hand but was slow afoot, had below average hand speed, and had no left hook to speak of. Joe would give him a tutorial on his signature punch, but first he hit the road with his new act and handed off the day-to-day supervision of Bobick to Futch, who told Looney upon taking a longer look at Duane: "I'm a trainer, not a magician. If a fighter doesn't have it, only God can help him." As Futch worked him up the ladder, careful not to place him in over his head yet eager to see what he had, Bobick continued to win, yet in an unimpressive manner that disappointed Joe. Of the thirteen opponents he had beaten since joining Frazier, none of them were more than journeymen, and only former Ali foil Chuck Wepner had any name recognition to the general public. After a proposed bout against Ali fell apart when Ali temporarily retired, Futch arranged for seventh-ranked Bobick to face third-ranked Ken Norton, whom Futch had guided to that upset victory over Ali in 1973 and with whom he had since parted ways. Although Frazier was against the pairing with his close friend Norton—perhaps holding out for a more lucrative Ali bout to reappear—Futch claimed that Bobick had come to a crossroads. He had to prove himself against someone of stature.

The bout was scheduled for May 11 at Madison Square Garden, with Norton set to earn $500,000 and Bobick $250,000. Down in Landover, Maryland, the thirty-five-year-old Ali said he was going to get down on his hands and knees and pray that Bobick beat Norton to set up an Ali-Bobick bout. (Ali was in Maryland for a title defense against Alfredo Evangelista, a fifteen-round unanimous decision that Howard Cosell called "an exercise in torpor not to be believed.") Ali rhapsodized: "The white boy against a big, bad nigger like me. I'll be talking. The Klan will be marching. The brothers will be dancing in the streets. The cash register will be ringing." But when Bobick stepped into the ring and placed his 38-0 record on the line against Norton, the big payday Ali envisioned vanished like a tendril of smoke in a gust of wind. Less than a minute into the bout, Norton clubbed Bobick with a half-dozen

overhand rights, then drilled him with a right uppercut to his
Adam's apple. Under the barrage of blows that followed, the defenseless Bobick fell to all fours. He was counted out at fifty-eight
seconds. Over at the Riverboat Club in the Empire State Building, preparations were still under way for the "win or lose" party,
where the entertainment would be provided by the Joe Frazier
Revue. Denise Menz remembered, "I was still handing out comp
tickets when the fight ended." Frazier consoled the woozy Bobick
in his corner before heading off to sing. Chagrined over the choice
of Norton as an opponent, he had no immediate comment to the
press on the outcome, saying only: "I've been quiet 'til now. So I'll
stay quiet." But Ali said what Frazier was surely thinking: "Bobick
lost a chance for ten million dollars in that one round."

As the unrealized expectations of Bobick circled the drain in
the ensuing years, during which he eventually cut his ties to Frazier
before retiring in 1979, Marvis set a course for the 1980 Olympic
Games in Moscow. Under the supervision of Benton, head trainer
Val Colbert, and amateur coach Sam Hickman, Marvis had his first
amateur bout in March 1977 at the Blue Horizon in Philadelphia,
where he beat an opponent who was sixty-six pounds heavier en
route to the Pennsylvania Golden Gloves Novice Heavyweight
Championship. From there, he won the state Golden Gloves Open
Heavyweight Championship the following year. He won it again in
1979 by planting one Ed Bednarik on the canvas with a right hand
at 2:59 of the first round. Hochman reported: "Bednarik went splat,
like a load of wet plaster." Pennsylvania State Athletic Commission
Chairman Howard McCall projected stardom for young Frazier
as a pro, assuring the only way he would not earn big money was
"if they stopped printing it." That year, he also won the Golden
Gloves Tournament of Champions in Indianapolis and the World
Junior Amateur Championship in Japan.

Early in the morning as he did his roadwork with light middleweight James Shuler, Marvis would occasionally holler: "The

winnah and newwwww champeen of the world . . . Marvissss Frazzzzzier!" By early 1980, Benton had shaped him into what his cousin and later stablemate Rodney Frazier called a "beautiful boxer." Futch extolled his "good left hand," of which he added: "He gets inside, he rips 'em off just like his daddy." With each victory, his daddy was increasingly impressed, exclaiming with a big smile: "The chip did not fall far from the stump." As the star attraction of the Joe Frazier Amateur Boxing Club, Marvis won forty-four consecutive bouts before he was beaten at the Summit in February 1980 by Tony Tubbs, who represented the Muhammad Ali Amateur Boxing Club and whom he had outpointed in a split decision the year before in team action at Resorts International Hotel Casino in Atlantic City. Only weeks later, he had an invitation to join the U.S. Amateur Boxing Team on a trip to Poland, only to decline when his father had a dire premonition. Subsequently, fourteen boxers and eight others accompanying the team perished when their jet crashed short of the runway in Warsaw.

Joe told me years later, "I told him not to go. Two weeks before that plane crash, I had a dream of a big fire. My whole family was burned up. All of them—gone. It was a house, not a plane, but I just had a bad feeling."

World events conspired to deny Marvis the Olympic gold medal that had been his guiding star. On March 21, 1980, President Jimmy Carter announced the U.S. boycott of the Olympic Games in Moscow in protest of the Soviet Union's 1979 invasion of Afghanistan. Marvis won the National AAU Championship in Las Vegas in May and, in the slim hope that the boycott would be withdrawn, reported to the U.S. Olympic boxing trials in Atlanta. There, he won a closely contested decision over Mitch Green in the quarterfinals and moved on to the semifinals to face James Broad, who had won the All-Army Interservice and World Military Championship. Eleven seconds into the bout, the 215-pound Broad tagged Marvis with a right hand that traveled no more than six inches and toppled him to the deck. He was counted out at twenty-one seconds of the first round.

For a few terrifying moments, Marvis just lay there on the canvas. Though he would later say he was still conscious, he could not move his arms or legs. As he recovered, he was helped to his stool and then into an examination room, where he was seen to by an Atlanta neurosurgeon. According to Dr. Donald F. Grady, Marvis had come "very, very close" to ending up like Darryl Stingley, the New England Patriots wide receiver who was hit in a 1978 exhibition game and permanently paralyzed from the neck down. But Dr. Grady diagnosed a pinched nerve and told Marvis he would be fine. "I know exactly how that feels," Joe told Ed Hinton, reporting for the *Philadelphia Daily News*. "When you get hit like that, it pops your head back and jams your neck into your spinal column, and it paralyzes the whole body." Joe said he was not worried about it.

He then looked at Marvis and said, "Come on, son, let's go home."

Cars were more than just a lavish expression of wealth to Joe. They spoke to his desire to get up and go and were a remedy for the devouring restlessness that he found so unbearable. Even before he had the wherewithal to accrue the fleet of luxury vehicles that sat in his driveway, which Denise Menz encouraged and helped him to parlay into a limousine service to offset some of the expenses related to the gym, he loved popping open the hood and tinkering with them. One could even say that fixing automobiles was something of an obsession, given how he would stop to help fellow motorists who were stranded alongside the road. Scattered across Philadelphia and beyond were countless people who had found themselves looking on in astonishment as the former heavyweight champion got the jack out of his trunk and began changing their flat. Burt Watson, once his business manager, said that it was common for the two of them to be driving to an event when the Samaritan in Frazier would come out. As Joe hopped out from behind the wheel and investigated the problem, Watson would sit in the car just shaking his head, thinking: "Oh no, here we go again." When it came to the upkeep of his own cars, it was not beyond him to a scour the junkyards of Philadelphia in search of an elusive part.

The junkyards in Philadelphia were hooked up to an intercom system called the "Hoot and Holler," which enabled them to work together to locate parts for customers. Upon hearing on it that Joe was in the area and looking for a differential for a '69 Caddy—in shorthand, "a rear"—young Michael Averona immediately informed his father, Sonny, who told him: "Get in your car and go get him!" Michael jumped in his car and began a search of Southwest Philadelphia, shooting up Passyunk Avenue and onto Essington Avenue until he spotted Frazier on the other side of the street driving a black limousine. Quickly, Michael pulled a U-turn. Then he pulled up alongside Joe and began shouting to get his attention. But Joe had the radio on with the volume up and did not hear him. So Michael swerved closer to him. Joe looked over at him in alarm. Michael yelled, "Pull over! Pull over!" When Joe did so, Michael pulled up in front of him, blocking his way to keep him from speeding off. Michael then hopped out of his car and approached Joe, seated behind the wheel in a cowboy hat. Michael said, "Joe, I hear you need a rear for your limousine. My father sent me to get you—Sonny Averona. We have one for you."

Joe followed him to the junkyard. There, Michael introduced him to his father, who told Joe: "We have the rear for your car. Back your car up to the gate." Michael would remember, "Shoulder to shoulder, Joe was gigantic. He was like a friggin' rock." To pay for the part, Joe produced a knot of hundred-dollar bills that Michael said "had to be three to four inches thick." But Sonny was not having it. He told him "the rear" was on him, enthusing: "I loved you as a champion and what you did for boxing." When Joe pressed his willingness to pay, Sonny told him, more firmly now: "The rear is nothing, Joe." Joe looked over at Michael and said, "Come on, get in the car." With the differential stowed in the trunk, Joe pointed the limousine back to North Philadelphia and drove him to the gym. Inside, he told the young man: "Pick anything you want."

"What?" Michael asked in disbelief.

"Go on," Joe said. "I'm gonna give it to you."

Michael would say years later, "Right off the bat, I knew what I wanted—this big heavy bag. Brown leather, the biggest one you could get. It wouldn't move when you hit it. Bang! It was like hitting a cinder block wall. He said, 'Ahhh, I train on that one.' He pointed to another one: same bag but smaller. He said, 'How would you like that one instead?' I said, 'I love it.'" One of his guys unhooked it and he autographed it."

The junkyard owned by Sonny Averona became a favorite hangout for Joe, a place where he could kick back and swap stories. "Every day," Michael said years later. "He was down here every day." Sonny became not just a friend with whom Joe would come to share occasional holiday dinners but a partner onstage as the years passed. Along with running the junkyard, Sonny fashioned a singing career when he was well into his forties by doing a lounge act in tribute to Frank Sinatra, which Sonny advertised as "the sound of Sinatra with the Averona style." Until his death at age fifty-five, in 1992—at which point Michael inherited the show, which he expanded to include a tip of his hat to Dean Martin—Sonny starred on the casino circuit in Atlantic City and Las Vegas. When he asked Joe to join him onstage, he did not have to ask him twice. Together, they appeared at the popular South Philadelphia spot Palumbo's and the Claridge Hotel & Casino in Atlantic City, among other venues. Michael Averona would say, "When Joe stepped onstage, you paid attention."

With Marvis up and running as an amateur, Joe revisited his love affair with music. Hard work had enabled him to overcome his physical shortcomings to ascend to the pinnacle of boxing, and he was certain that the same hard work would help him to overcome his deficiencies as a singer and forge a second career. Carlo Menotti had given him some valuable pointers, yet his breath remained too short, his range too narrow, and his phrasing too imprecise. But with the announcement of his retirement, he recommitted himself to showbiz, giving it his full attention. With Florence and their five children back in Whitemarsh Township, he moved into a penthouse apartment in Manhattan and opened an office. To sharpen his skills, he worked for two or more hours a day with Eddie Jones, the jazz double bassist who played with the Count Basie Orchestra in the 1950s. Of his sessions with Jones, Frazier said: "Over and over. Up and down. Hit the wrong note, do it over." Jones told him: Be yourself. Not everyone can be Sam Cooke or Nat King Cole. Jones helped him assemble his group of backup performers: three disco dancers, a trio of pop singers called "Smoke," and an eight-piece band. Stan Hochman attended their November 1976 debut performance at the Northeast Hilton in Trevose, Pennsylvania. Not an easy critic, Hochman came away impressed, writing in the *Philadelphia Daily News*: "Frazier has made incredible strides as a singer. He is twice the performer he was four years ago, slicker, smoother, more in control. If only his voice was a little mellower, a touch richer . . ."

Crowds were so big the following February at the Rainbow Grill, on the sixty-fifth floor of the RCA Building in Rockefeller Center, that he was held over for a second week. He performed two shows nightly—9:15 and 11:30—and altered his offerings for the second set in the event someone stayed on. Striding onstage in a silver-and-black warm-up robe—which he stripped off to reveal a tux with sparkling lapels—he planted his feet, turned to the audience, and unpacked his repertoire, including "Knock on Wood," "When Something Is Wrong with My Baby," "Proud Mary," and—as always—"My Way." The crowd loved him. Syndicated columnist Earl Wilson enthused, "He kayoed every song." In an interview with Wilson in advance of his opening performance, Frazier said ebulliently: "Can't get nobody to sleep with me anymore, 'cause I sing in bed all night. I sing ridin' in the car and get stopped by the police." He then took a playful poke at Ali, asking Wilson: "Can you imagine Muhammad Ali tryin' to sing 'Proud Mary'?" At the end of his second set, Joe was so pumped up that instead of taking a car back to his apartment, he would get into his workout clothes, ride the elevator down to the ground floor, and run up Fifth Avenue and back down Eighth Avenue before heading home.

Administrative responsibility for the band fell to Denise, who had come to New York to live with Joe. With Frazier footing the bill and fourteen or more personnel to pay and keep track of, his music career quickly became what she called "an organizational nightmare." Agents lined up bookings for the group at small venues up and down the Eastern Seaboard and beyond, some of which Denise said had to be "burned" because they simply did not pay enough to break even. "Our performers were not what you would call top-tier professionals," said Menz. "It was like babysitting a kindergarten class. Who needs a taxi to the airport? Who needs ten or twenty dollars for this or for that? They even stole pillows from the hotel! And then there were the costumes to see to. It never ended." Money that came in from the show dates was used to cover expenses, which included a weekly payroll of more than forty-five hundred dollars. Down in Philadelphia, Bruce Wright fielded the bills that came in and grew concerned, asking, "Denise, what in the world is going on up there?" Given the overhead that running a band entailed, Menz contended Joe would have been far better off if he had confined himself to personal appearances and commercials. For a Miller Lite commercial in 1978, he was paid twenty-five thousand dollars plus royalties for a single day of work. "Joe had to sign a paper with Miller Lite that he was retired, because they only used retired athletes for those spots," said Menz. "I remember we did forty-five takes." The spot aired in 1978 and revealed a droll Joe: In a tuxedo and big red bow tie, he boogies into a swanky saloon with his backup singers on his heels, snapping his fingers and singing a jingle for the beer, only to finish and encounter stone silence by the other patrons until he gives them a menacing frown. Only then do they erupt in cheers, at which point Joe dissolves into a satisfied smile.

Tension began to grow between Joe and Denise. Even though she never wavered in her affection for Joe, she found herself becoming increasingly overwhelmed by the pressures of the band. Joe himself began to discover that he had been spoiled by the more or less dependable way business had been conducted in boxing, where his contracts were typically letter-perfect and the money that was due him was always placed in escrow. Show business operated in a far more haphazard way that invited an array of unexpected snafus. Whenever Joe would become irritable, Denise found herself bearing an ever-larger burden. Borrowing a phrase from James Brown, "The Godfather of Soul," Joe would remind her with cutting brusqueness: "I pay the cost to be the boss." Annoyed, Denise would think: Yes, you do. You do pay the cost, but I am dying here. She would remember years later: "Along with the problems with the band, we always had trouble with other women. They would come up to me and ask me to hook them up with him. And the party life. He could never get enough of it." As she weighed what to do, she remembered what her father had always told her: "Stand on your own two feet, Denise." Eventually, she gave up her apartment in Philadelphia and moved back home to Rio Grande, New Jersey, where she and her brother Jay owned a restaurant. She remembered, "I did it for self-preservation."

Even as he applied himself to his singing career with vigor, Joe could not bring himself to let go of boxing—or Ali. When Ali said offhandedly in an interview that he would consider a fourth match with him, Frazier seized upon it. "You know I ain't got no rabbit's blood in me," Joe told the Associated Press in August 1977. "If a guy wants to take me on again, you know I'm not gonna run." Family and friends shuddered. Only fourteen months had elapsed since Joe had staggered from the ring after Foreman throttled him a second time. To get his comeback under way, he began training in September and advised Bruce Wright to begin lining up an opponent. With Joe in the office, Wright called Eddie Futch in California and placed him on the speaker. For a half an hour, Futch found that as hard as he tried to dissuade him, Frazier only dug in his heels deeper. Finally, Joe said: "Do me a favor. Come back and take a look." On October 17 and 18, 1977, Futch stopped in at the gym and did just that. To his surprise, he was impressed by what he saw. Joe appeared to have his old zest back and looked better than he had before the second Foreman bout. Though he told Frazier that he would prefer that he reconsider and stay retired, he agreed to work with him again, telling *Philadelphia Inquirer* writer Skip Myslenski: "Fighters are the last ones to realize the truth. So . . . so, what can you say?"

Eleven days after Futch performed his inspection, Ali faced Earnie Shavers at Madison Square Garden in a bout that stripped yet another layer of dignity from the hallowed champion. Ali was awarded a unanimous fifteen-round decision, yet it was a long and grueling evening for him, as the hard-hitting Shavers nearly knocked him out in the final round. Sitting at his home in South Jersey, then-fourteen-year-old John DiSanto found himself troubled by the beating Ali took, and how he seemed to age more with each passing round. "Grease from his face had gotten into his hair," DiSanto remembered. "As the fight wore on, it appeared as if his hair had turned gray before our very eyes." Toward the end of November, the young boxing fan pleaded with his parents to take him to see Frazier at a performance at the Woodbine Inn in Pennsauken, New Jersey. With the mournful image of Ali still front and center in his mind, DiSanto sat at the bar over a soda and jotted down a note for Joe on a piece of paper. The boy wrote: *Dear Joe. I'm a big fan. I've followed your career. You're a hero of mine. I read or heard that you are thinking about making a comeback. I would hate to see that because you've had a great career and there's no reason to take that risk.* With that, DiSanto folded the note and asked the bartender to deliver it to Joe backstage. DiSanto would remember, "They called me over later and introduced me to him. I shook his hand. But I have no idea if he read or even received the note."

Increasingly, Frazier had come to believe that he had retired too soon. Seeing Ali get dragged through fifteen inglorious rounds by the inexperienced Leon Spinks in February 1978 and lose his world heavyweight championship in a split decision only confirmed that. Even though Spinks had won the 1976 Olympic gold medal, he had entered the ring with only seven pro bouts. That Spinks had been trained by Sam Solomon and George Benton in his own gym had to have been even more vexing to Frazier, who told the press: "With the guys out there now, why shouldn't I take another shot?" But there would be obstacles. For a potential Shavers bout, Frazier set his price at $1.1 million, well above the $750,000 that was on the table. When television balked at underwriting the promotion, it was scrapped. Teddy Brenner signed Joe to a contract for a potential bout with Scott LeDoux at Madison Square Garden on April 16, 1978, but he could not line up LeDoux, who had fought Spinks to a draw a year before and was now hoping to position himself for a rematch. While Brenner searched for a replacement—Jimmy Young? Stan Ward? Mike Weaver?—Frazier signed to take on Gerrie Coetzee in Johannesburg. For Frazier, it shaped up as an attractive payday—$460,000 plus another $170,000 in endorsements—yet it had problematic overtones. South Africa was then in the stranglehold of apartheid and had been banned from the Olympic Games since 1964.

For an African American who had grown up in the Jim Crow South, and who had cultivated relationships with Frank Rizzo and others in the white ruling class, it could not have been a more ill-considered play. While Futch had gone to South Africa with Bob Foster for a bout against Pierre Fourie and later would accompany others, the Coetzee offer came at the height of the global anti-apartheid crusade. Although Frazier demanded that the seating for the bout be integrated, as it had been for the Foster-Fourie fight, it did not quell the outcry that erupted. The NAACP sent Frazier a telegram that Bruce Wright said "explicitly asked him not to go to South Africa, the plea being based on indignities to the black race that existed there." Ultimately, the deal came apart when the Transvaal Provincial Boxing Board of South Africa vetoed the bout, reportedly due to a clause in the contract that sheltered Frazier from a 6 percent tax obligation. As South African promoter Hal Tucker worked behind the scenes to iron out the problem, Joe withdrew from the arrangement, and a chagrined Wright lamented that it was "reasonable to say that Joe kicked away seven hundred thousand dollars by not going." *Philadelphia Daily News* columnist Chuck Stone excoriated him, observing: "Golly gee, think of how badly all those South Africans languishing in political prisons must feel. . . . No comeback trail should be littered with the tortured and murdered bodies of South Africans."

Twelve years later, in an event at the United Nations, Frazier would meet the noblest of those captives, Nelson Mandela, a former amateur boxer who served twenty-seven years in jail as a political prisoner. Joe would give him his jewel-encrusted championship belt that day in 1990, saying: "Mr. Mandela, you deserve this. This is from my heart." But while Joe was still on the comeback trail, Stan Hochman ventured that "you could get 3–1 that Frazier couldn't tell you what apartheid was." With Coetzee now out, Frazier accepted a bout with Kallie Knoetze. Of all things, Knoetze was a South African police detective, which drew a howling objection from Don King—the very same Don King who divvied up swag with some of the worst authoritarian cutthroats on the planet. "Fighting a South

African is helping those who support white supremacy," boomed King—who would himself promote Coetzee years later. The bout was scheduled for May 14 in Las Vegas but had to be pushed back when Frazier came down with what appeared to be the flu. Only days later it was discovered that he had contracted acute viral hepatitis and would be sidelined indefinitely. Frazier said, "The Good Man must be trying to tell me something." Atop a column by Dave Anderson in the *New York Times* on the day Joe had been scheduled to face Knoetze, the headline read: HEPATITIS DID FRAZIER A FAVOR.

———

Few fighters have ever looked upon retirement as the pleasurable escape that Archie Moore once envisioned it to be. The bewhiskered "Old Mongoose" used to say when he exited the ring at the end of his long career, he would hang his gloves way up high, burn his workout clothes, and eat whenever and whatever he pleased. He would stay up late and "inhale the fumes of good jazz in smoky nightclubs," have his wife hide his scrapbooks and photo albums, and go over to the gym each day. There, he would shoot the breeze with the other old-timers and, as he cast an appraising eye on the youngsters in the ring, he would tell himself that he could have beaten any of them in his day. For Joe Frazier, letting go would not be so easy.

Only thirty-four when hepatitis derailed his comeback, Frazier was old for boxing yet still in his prime, at a point in life when successful men in other fields are able to look forward to decades of productivity. Although he had the gym, occasional singing dates, the limousine service, and later a restaurant, none of it provided him the thrill that he derived from fighting, of which he once said: "In the ring you can get rough, rugged. I like that. I get a kick out of that." For the action that he always craved, he stopped in at the casino that had just opened in Atlantic City, and not a day passed when he did not have a handful of lottery tickets jammed in his pocket. To pass the hours, he would stop by the junkyard, work on one of his cars, and sit with Sonny Averona. On some days he would set up a projector and go over films of his old bouts, which was what his niece Dannette found him doing when she unexpectedly stopped by. For hours they sat there together as his history unfolded before their eyes: George Chuvalo, both Bonavenas, Buster Mathis, Jimmy Ellis, both Quarrys, both Foremans, all three Alis—all the victories and defeats. Dannette would remember, "We sat there so long we had to send out for food." How far away those years now seemed.

Depression set in. Even if Joe could not bring himself to admit it, it showed up in his behavior. Heated words were exchanged with Bruce Wright over spending and expense practices, which had ushered in tax problems. When Joe fired him the next day, Wright would remember that "of all the things I said about him, the only one that upset him was when I called him old." Stan Hochman stopped by to see him early one day at the gym and became alarmed at the condition he found him in. Wearing a jacket with the words ONE OF GOD'S MEN stitched on the back, he had two handguns shoved under his belt and smelled of alcohol. Unaware that Denise had moved back home, Hochman called her, told her what he had seen, and said, "As a sportswriter, I never get too close to the athletes I write about. But I have a special place in my heart for Joe and am hoping you can talk to him." Denise told him they had split up and added sadly: "Stan, he would not listen to me anyway." Hochman paused and replied, "Ah, I see. I am sorry to hear that." Hochman would subsequently write: "He has become stubborn, moody, a poor listener. He has surrounded himself with people who nod and say, 'You can do it, Joe. You can do it, baby.'"

The *Philadelphia Evening Bulletin* carried the reassuring headline in June 1980: FINANCIALLY, FRAZIER STILL A HEAVYWEIGHT. Perhaps, yet it seemed as if he was working his way down to a lightweight with abandon. Day by day, "the love" just dribbled away on this and that. For years, the gym had been bleeding him for five thousand dollars a month in upkeep and utilities. The Joe Frazier Revue had been what Hochman called "a six-figure wipeout." Under the supervision of his brother Tommy, the limousine service was scarcely a break-even proposition. Along with the cost of running his house in Whitemarsh Township, he had additional expenses connected with the South Carolina plantation, which was occupied by his mother and two sisters. Moreover, there were his five children with Florence to care for and child support for a son—Joseph—he had in September 1980 from a relationship with real estate agent Joan Mahoney; Frazier had offered child support for the two children in New York that he had with Rosetta, but she rejected it. Off the books, there was his fondness for blackjack, which he played heavily and not well. According to Ali aide Gene Kilroy, "Joe was the worst blackjack player in the world. He would split picture cards." Although income came via a seventy-thousand-dollar annual pension Wright had set up, personal appearances and commercials—along with Miller Lite, he had done ads for Mennen Skin Bracer and Blue Bonnet Margarine—he launched yet another comeback. Jokingly, Joe said he had to "pay the rent."

Boxing does not protect its aging stars from themselves. For Ali—perhaps its greatest star—it would be a long and painful goodbye. In September 1978, with Frazier on hand to sing the National Anthem at the New Orleans Superdome, Ali captured the title back from Spinks in a unanimous fifteen-round decision to become the first man to win the world heavyweight championship three times. Although Ali surrendered it when he yet again announced his retirement in June 1979, he came back just over a year later to challenge Larry Holmes for the WBC portion of the crown. In a ring set up in the parking lot at Caesars Palace in Las Vegas, the far younger Holmes pummeled a more or less defenseless Ali in the blazing heat for ten horrifying rounds. In his book *Ali: A Life*, author Jonathan Eig observed that as Ali was outpunched 340 to 42, his manager Herbert Muhammad "sat at ringside with his head down, unable to watch yet unwilling to stop it." Only when Angelo Dundee caught his eye at the end of the tenth round did Herbert give a brief nod for him to stop the slaughter. In December 1981, eight days before Ali dragged his thirty-nine-year-old body through a final beating, at the hands of Trevor Berbick, in the Bahamas, Frazier had an appointment in Chicago for what would be his own sad denouement.

Five and a half years had elapsed since Foreman had beaten Joe on Long Island. With Wright removed from the picture, Joe handed his business affairs over to Sharon Hatch, who had once worked at Madison Square Garden and with whom he would have a relationship; they had a son together in December 1982— Joseph Jordan Frazier. The opponent Hatch lined up was an ex-con, Floyd "Jumbo" Cummings. While Hatch would then say that Joe was not in any urgent need of the eighty thousand dollars he would receive from promoter Bill Cooley—Cummings was paid ten thousand dollars—George Benton said, "Why else would he be doing this except for money?" Benton would not accompany Joe to Chicago. Nor would Eddie Futch, who said it saddened him to see Joe come back. Futch told Lewis Freedman of the *Philadelphia Inquirer:* "The legs start to go. You see punches and adjust a split second too late. You see openings just a split second too late. Age does dim the reflexes. There are no exceptions." The evening before he was to leave for Chicago, Joe drove down to Rio Grande to pay a visit to Denise. When Joe attempted to break the ice with small talk, asking her what she thought of the Cummings fight, she gave him the cold shoulder. "If I had any say in it, that fight would never have happened," she later said. "It was beneath him."

"Guess what kind of bird don't fly?" Joe asked at the mock weigh-in the day before the bout. From nearby, nephew Rodney Frazier squawked: "A jailbird!" Cummings had served twelve years of a seventy-five-year sentence at Stateville Correctional Center in Crest Hill, Illinois, on a murder conviction. Only sixteen years old and just up from his birthplace of Ruleville, Mississippi, he had been the wheelman in a grocery store robbery with five accomplices. One of them shot and killed the grocer. Cummings, known as "the barn boss" in stir because of his unchecked appetite for brawling, spent the better part of his early days in prison in the isolation cell, only to shape up when the prison started a boxing program. Jumbo said, "Boxing was a natural for me 'cause all it is, is street fighting with a little polish." Eight years younger than Frazier and two years out of prison, Cummings packed 223½ pounds in a six-foot-two sculptured physique and had a 17-1 record. Fearful not that he would hurt Frazier but that he would kill him in their ten-round bout, Jumbo said he planned to bring along his two attorneys in case he had to appear at an arraignment.

Forty days shy of his thirty-eighth birthday, Joe tipped the scale at 229 pounds, the highest he had ever weighed for a bout. To dispel any concern that Frazier was unsound, the Illinois Athletic Commission had promised to have him undergo a physical examination. When they settled for looking at records supplied by a Philadelphia doctor who had cared for Joe, Hochman creamed them in his column, asking: "Who are these guys? The same people who sold tickets to The Who concert in Cincinnati?" Unconcerned, commissioner Bob Goodsitt said with a shrug: "An X-ray is an X-ray, a doctor is a doctor." None of the three networks would get near it. CBS boxing advisor Mort Sharnik told Frazier personally: "Joe, my memories are sacrosanct, and you should be treated like a national treasure. You put on the greatest sports event in history and when you retired five years ago, I agreed wholeheartedly. In the five years since, nothing has caused me to change my mind." On the day of the bout, Joe received a telephone call from Ali in the Bahamas. The two proceeded to cook up plans.

"We gotta make all the old men proud," said Ali.

"I hear you," said Joe. "I'm gonna hold my end of the deal up."

"I'm gonna do my best down here, too," said Ali.

"I don't want to hear none of that I'm-gonna-do-my-best stuff," said Joe. "I'm talkin' about you holdin' your end up."

"Yeah, we're old men and we gotta show the world we can do it," said Ali.

"Don't call me old," Joe said.

Situated across the street from the Union Stock Yards—which had closed in 1971 yet still somehow seemed to discharge a foul odor—the archaic International Amphitheatre had to seem a million miles away from Madison Square Garden to Joe. With a crowd of sixty-eight hundred squarely in his corner—"Boogie, Joe! Boogie, Joe!"—the man who once wore a bloody slaughterhouse apron himself charged from his corner at the bell and began throwing what had once been his Hall of Fame left hook. He missed. Then missed again. Even when Joe connected with it, it did not appear to faze Cummings, who stood back and sneered. Scoring with right uppercuts yet breathing hard by the third round, Cummings had his way in the early rounds, while Frazier seemed to have ten-pound weights attached to his ankles. Seated at ringside, John Condon, his old friend from the Garden, was overheard by *Chicago Tribune* columnist Steve Daley: "No legs. Joe is going to have to get awfully lucky."

Had he ever stooped to bring on Jumbo Cummings as a sparring partner back in his prime, Joe certainly would have hospitalized him. But on this evening, still possessed by gallant pride but with none of the untamed savagery that used to accompany it, Frazier was indeed too old and too slow and too heavy, his once-hard abdomen now jellied with a layer of flab. Working the corner along with Val Colbert instead of Futch and Benton, Marvis Frazier urged his father on between rounds. Frazier marched out and won the fifth, landing a solid left hook that seemed to jar Jumbo, but only briefly. "Atta boy, Joe!" shouted someone in the crowd, now beginning to lean to Cummings. The erstwhile "barn boss" pinned Joe in a neutral corner in the eighth and hit him at will as

Frazier stood there unable to move. Blood poured from his cut lip, both eyes now beginning to swell. Only guile enabled him to escape back into the safe harbor of ring center. As the bout came to a weary finish two rounds later, it appeared that Joe would go home with the fourth loss of his career against thirty-two victories.

The judges called it a draw. Boos rained down on them.

Jumbo slammed his glove on his forehead in disbelief.

"You know I was robbed, man," he said later at his press conference. "Joe Frazier is a great fighter, but he knows in his heart I won."

Frazier said it could have gone either way. Then, he added: "Most people think I'm an old man, but I don't think you saw an old man out there tonight."

Acid poured from the computer terminals on press row. Hysterically, Cooper Rollow shouted in the *Chicago Tribune:* "It may have been the worst decision in the history of professional boxing." In a more minor key yet no less critical, *Chicago Sun-Times* columnist John Schulian observed: "Jumbo Cummings cried a river when the news hit him harder than the forlorn ex-heavyweight champion had in ten rounds, and a sweating, stinking mob belched up its anger with him. But the point that should be made in the haunted aftermath has nothing to do with justice in a sport that never had any to start with. The point that should be made is that Joe Frazier never should have been out there in the first place." Down in the Bahamas for the upcoming Ali-Berbick bout, *Philadelphia Inquirer* columnist Bill Lyon worried that the unjust draw would only encourage Joe to come back yet again, perhaps even with the hope of pairing up with Ali a fourth time. Lyon wrote: "No, no, a thousand times no!"

To the relief of those who loved and admired him, there would not be another bout, not with Ali or anyone else. Contrary to what Joe would say in the days that followed, he was finished when he exited the International Amphitheatre into the cold and rainy

Chicago night. Only days later, it would end for his enduring rival in Nassau, Bahamas, where Ali was beaten in a ten-round unanimous decision by the plodding Berbick in a setting just as unseemly. Far older in appearance than his thirty-nine years, one of the greatest champions of any era closed out his career in a ring that had been set up on a high school baseball diamond. A cowbell was used to signal the beginning and end of each round, which only underscored the words uttered by Gypsy Joe Harris years before and now applied to both Ali and Frazier: once hammers, they were now nails.

CHAPTER ELEVEN

SONS

The Fighting Fraziers (L to R): Hector, Rodney, Joe, Marvis, and Mark, 1983. Getty Images

Gypsy Joe Harris stood on the Ben Franklin Bridge and looked down at the churning currents. For years, he had traveled by foot over the walkway on the bridge from his boyhood home in Camden to Philadelphia and back again. Standing at the apex of the span, he had a panoramic view of both cities: Camden, its shoreline scattered with warehouses and water towers, and Philadelphia, its stunted skyline then still no higher than the statue of William Penn atop City Hall. One hundred thirty feet below, the Delaware River flowed between the two banks and under the bridge. To the occupants of the cars and trucks that whizzed by, the forlorn figure who stood at the railing would have been no more than a blur that passed by in the blink of an eye.

Some years had elapsed since the Pennsylvania State Athletic Commission had discovered Gypsy Joe was blind in one eye and stripped him of his boxing license. Certain that they knew of his condition all along and had overlooked it, Gypsy could not help but view his ouster as a cruel irony: By purportedly saving him from the perils of the ring, they consigned him to an even worse outcome on the street. With no skills to call upon other than boxing, he ended up in the stranglehold of drugs and alcohol. "Heroin is just like sugar, man," he told *Philadelphia Inquirer* reporter Robert Seltzer years later. "It relaxes you. It takes all your problems away." Only when friends began dying off did he seek hospitalization, yet the treatment he received only addressed his cravings, not the aching hollowness that would lead him up on that bridge. As he stood there and weighed his alternatives, it occurred to him that years before he had won "a little medal" for swimming at a recreation center as a boy.

"I told myself I was going to jump," said Gypsy Joe. "But then I thought, 'Hey, I can swim!' It seemed kind of funny later. Here I was, I wanted to drown myself, and could swim."

Chances are he would not have survived that fall any better than the one that claimed his boxing career. For years, exactly what happened remained a puzzle. One theory that has circulated is that Gypsy Joe, overshadowed by Joe Frazier and upset by the lack of attention he had been receiving from Yank Durham, was planning to leave Durham and sign with Bernard Pollack, a wealthy mink farmer and furrier who owned the property Ali later converted into his training facility in Deer Lake. Gypsy Joe had befriended Pollack and had been a frequent visitor at the farm, where the two would take long walks and talk; Pollack also was blind in one eye. According to Gypsy Joe's younger brother Anthony Molock, Gypsy and Durham had a "big blow-up" over it at the Twenty-Third PAL prior to his scheduled bout against Manny Gonzalez on October 14, 1968. "Gypsy stormed off at the end of it," said Molock. "Next thing you know, they 'discovered' he had one eye." Gypsy Joe had become his own worst enemy even in the eyes of those who cared for him, and he believed that either Durham, his associate Willie Reddish, or promoter Herman Taylor had dropped a dime on him. Even Molock could see where Durham or the others had a point (if indeed they did it): "All the trouble Gypsy was? And now he was up and leaving? *I* would have dropped a dime on him."

Aware of the eye problems Frazier had as the 1970s unfolded, Gypsy Joe had to feel as if he had gotten shorted. When his appeal to have his boxing license reinstated was denied in 1972, his drug addiction spiraled out of control. He told Seltzer that "me and a buddy did twenty-five bags together. It kept us high for days. . . . You nod away." Some speculated that while he was up at Deer Lake, he worked as a spy for Frazier yet also slipped information on Joe to Ali for some extra money. An insider told me, "Could be. The shape Gypsy was in, he would have sold out Jesus Christ himself for a fifth of whiskey." When Gypsy Joe stopped by the Cloverlay Gym, he would spot Frazier and shout, "Hey, Joe Frazier! Gimme some of that money you have!" Joe gave him a job instead. But just as he had done during his career, Gypsy Joe showed up only when the spirit moved him. When he did come in for work, he was often late. In the biography he authored, *Gypsy Joe: Son of Philadelphia*, Molock said that Frazier summoned Harris to his upstairs office to discuss his poor work attendance. Frazier told him, "This is a business, Gyp. You understand?" Gypsy Joe said he would do better yet continued to show up late or not at all. When Frazier smelled alcohol on his breath, he again called him to heel.

"Gyp, I know what you're doin' and where you're goin'," Frazier told him. "Remember, I live in North Philly, too. You need to get some help."

"Help for what?" Gypsy Joe replied. "The only help I need is money! Y'all made millions! I saw it in *Time* magazine y'all made $2.5 million! I get paid sixty dollars a day and now you hasslin' me about what I do on my time."

Frazier shot back, "I don't give a shit what you do on your own time! It's what you ain't doin' here is what I care about. You drinkin' and partyin' all night and can't get here to do your job. That's what I'm talkin' about! Gypsy, from what I hear, you need to get help!"

"Well, help me then!"

"Damn, Gyp! I gave you a job and I been payin' you for nothing lately. What you gonna do about your own situation?"

"That the way you feel? Then fuck you and your job."

Drugs frightened Frazier in a way no man ever did in the ring. Although he was no stranger to the party scene and drank himself, increasingly so once his career had come to an end, he gave drugs a wide berth and always worried that one of his children would get caught up in them. With an incredibly high pain threshold, he even waved off painkillers whenever he suffered a cut and there were stitches to be endured. To prevent colds and other conditions, he relied on a concoction that had been favored during his youth in Beaufort. He would take Courvoisier, cod liver oil, lemon or some other citrus peel, and rock candy (or Hall's menthol lozenges if he could not get his hands on rock candy), blend it together in a jug, and leave it to ferment for thirty days. When he could feel an illness coming on, he would reach for the jug and take a swig of what he called his "cure-all." Out of curiosity, his friends tried it when he offered it to them—once but never twice. "Good God!" said Denise Menz. "I told him, 'Keep that stuff away from me!'" But Joe abhorred drugs and often volunteered to speak out against them, once telling a school assembly in South Carolina: "Once you take something that will harm the body, you'll never be champ of anything."

Strung-out junkies were always on the prowl in North Philadelphia. One cold winter evening, Joe and Denise stopped at a nearby fried chicken place for a drive-through order in one of the limousines. A young woman no more than eighteen walked up to his window. Denise remembered, "She was so skinny, and she was shaking. She had on a sleeveless dress but no coat." The young junkie asked Joe for some money. He looked at her and replied, "Go home to your momma. Get off the street. You know what can happen to you out here?" Unsteadily, the young woman walked around the car. As Denise opened her purse and got out a bill, Joe flashed his eyes at her in anger, saying, "No! You give her that money, you know she'll buy dope with it." The young woman looked into the window at Denise and said, "Miss? Can you help me?" Denise jammed twenty dollars in her hand but told Frazier she had only given her five. As Frazier continued to vent his annoyance, Denise told him, "Maybe she'll use some of it to get something to eat." Denise would say years later: "Joe would have given her food if she had asked for that. He would have even driven her to see her momma. But under no circumstances was he ever going to give her a dime to buy drugs."

With each year that passed during the 1980s, drugs ate deeper into the neighborhood surrounding the gym. Crack vials were scattered along the soiled pavement. Razor wire was coiled atop high fences. Although he still had his home in Whitemarsh Township, the gym was now not just where Joe worked but had become a convenient crash pad. On friendly terms with the cops, he always kept the gym door open during the day and never had a problem other than the day he was outside working under the hood of one of the limousines. It was a Saturday, broad daylight, and there he was, wearing three diamond rings and a gold pendant that sparkled in the sun, when a large man jumped out of nowhere and tackled him in an apparent robbery attempt. Joe grabbed for the gun he had tucked under his belt, but the assailant got his hands on it first. He aimed it at Joe and yanked at the trigger. The gun jammed. As Joe wrestled him for it, the gun clattered to the ground, whereupon Joe picked it up. By now, a crowd had formed, and his brother Tommy raced out of the gym. As Joe stood back and took aim, Tommy leaped on his back and held his arms, shouting: "No! No! Think of Momma! Think of Momma!" The assailant ran off. Joe was uninjured except for cuts on his leg that had come from scraping it on a piece of fallen razor wire.

Gypsy Joe roamed these sad, sick streets until the end of his days. Never one to wear an eye patch over his dead eye, he remained the uncrowned champion of Philadelphia, still remembered in the pool halls and bars by those who were old enough to have once caught his colorful act. "Yo, Gypsy Joe!" they would cry. "Man, I used to see you at the Arena. You were something else." Occasionally, he would still wander in off the street and into the gym, where he would grab a chair and shout instructions to some young man up in the ring. "When you back up, keep throwing punches!" Frazier used to point to him and tell young fighters, "See that man over there? That man was the fightingest scamboogah to ever put on shoes." In and out of the hospital in the 1970s, Gypsy Joe cleaned himself up and got a janitorial job at City Hall that was set up for him by Mayor Rizzo. By the late 1980s, he had had three heart attacks and was subsisting on the $87.50 welfare check he received every two weeks. One morning when Joe showed up at the gym with his son Hector and two other fighters from an out-of-town trip, he found Gypsy Joe sleeping in the doorway.

"Whatcha doin', man?" Joe asked him.

"I'm hungry as a motherfucker," his friend replied.

Joe unlocked the gym door. "Come on in," he said, leading the way. "You got love?"

"I ain't got a fucking dime."

Joe eyed him and said, "Get a shower and we'll get you something to eat." Frazier told the three fighters to go around the corner and pick up some Kentucky Fried Chicken. Gypsy Joe took his shower and dressed in fresh clothes and sneakers Joe handed him.

When the fighters came back with the food, they all sat down at a table and tore into the bags. Gypsy grabbed a chicken thigh and began gnawing on it.

"I love you, Joe, man," he said.

"I love you, too," said Frazier, as he slipped some bills from his sock and folded them in his hand. "But you gotta do better, man. You gotta do better."

Gypsy Joe died of a fourth heart attack in March 1990. He was forty-four. Relatives chipped in enough to buy him a blue suit, an inexpensive coffin, and a $105 burial plot at Merion Memorial Park in Bala Cynwyd, where he was lowered in an unmarked public grave with two other people who had passed away that week. Sixteen years later, John DiSanto, who as a boy had written Frazier that cautionary letter at the Woodbine Inn, discovered that no headstone sat atop the grave and set to correct that by collecting two thousand dollars for a proper memorial. When it was done, it simply said: GYPSY JOE HARRIS 1945–1990. For anyone who remembered him in the ring, it did not have to say any more.

Five months before his untidy farewell against Jumbo Cummings in Chicago, Joe appeared with Marvis on the cover of *Sports Illustrated* under the headline: CHIP OFF THE OLD CHAMP? Less than a year before, on his twentieth birthday—September 12, 1980—Marvis had set out to answer that question when he launched his pro career at the Felt Forum in Madison Square Garden against Roger Troupe, a former wide receiver for the Philadelphia Bell in the World Football League. Under a one-year promotional contract with the Garden that paid him a fifty-thousand-dollar bonus, Marvis overcame some early jitters and acquitted himself well. Though Troupe buckled his knees at the end of the first round, Marvis walloped him with a left hook in the second round that had the scribes on press row wagging their pens. He followed up with a right hand that sent Troupe through the ropes. A round later and Marvis was standing with his gloves in the air, the winner by technical knockout 2:08 of the third round. Working the corner with George Benton and Val Colbert, Joe was not pleased by what he had seen. "You gotta breathe on him, son!" he had told Marvis between rounds. "Get close!" Later, Joe told the press, "Right now, we're in grammar school. Then you go to junior high, high school, and then college." Going forward, he would replace Benton and take charge of his boy himself.

History would have told him what a poor idea that was. Had Joe been paying close attention to complications that unfolded between the Quarrys—father Jack and sons Jerry and Mike—it would have acquainted him with the countless ways that the relationship between a father and son in the ring can go wrong. For a father, there is a line to walk between being too protective and not protective enough. Hard-nosed Jack Quarry exemplified the latter. Eddie Futch called him "the bravest manager in the world when it came to matches for his son Jerry." According to Futch, fathers tend to amplify the ability of their sons, who they "imbue with special powers, special qualities." From the perspective of the son, there are issues of ego, identity, and measuring up, perhaps some underlying pathology unique to the relationship itself. Perceptions can become clouded. Because money is also changing hands, it can become an even more convoluted partnership, with the potential for off-the-book arrangements and unfulfilled obligations. Nothing good has ever come of fathers taking over the ring careers of their sons.

Even as Marvis commenced his career with fanfare, there were problems at the gym that steeped Joe in gloom. To help curb his losses, Joe increased the gym dues for pros to a citywide-highest sixty dollars per month. Fighters picked up and went elsewhere. Some of the better young talent he developed, including 1984 Olympic gold medalists Tyrell Biggs and Meldrick Taylor, followed Benton out the door and signed with Lou Duva at Main

Events, where Biggs became a heavyweight contender and Taylor won the welterweight championship. Angrily, Frazier accused rival trainers of swooping into his gym and stealing fighters. Without naming names, he said that "today fighters are like whores. They go to whoever offers the biggest bankroll." Gone were the days when young fighters possessed the same loyalty that he had for Yank Durham. Stan Hochman wrote of Frazier in March 1982: "He stalks the gym like a brooding Othello, sensing Iagos in every corner. He calls anyone who disagrees with him a 'snake' and vows to get rid of them. . . . He forgets that a man chasing phantom snakes can beat the meadows down to parched stubble." But as the 1980s unfolded, boxing would become something of a family business, with Marvis followed into the gym by his half brother Hector and cousins Rodney and Mark; even Joe's boyhood friend Isaac Mitchell's son got in on the act, fighting under the name Tyrone Mitchell Frazier. Joe had management contracts with each.

The papers called them "The Fighting Fraziers." While Marvis lived at home with Florence and his sisters and would later marry, Hector, Rodney, and Mark stayed at the gym with Uncle Billy, where they trained, ate, and slept. None of them would go on to have more than an average pro career, yet Rodney remembered that the experience afforded him a chance to develop a close relationship to his uncle Billy, who Rodney would later say possessed three sides: "He could be loving. He could be compassionate. And he could be a tyrant." Rodney would say that while he was "generally a good guy, he could snap and become very accusatory of people." But Rodney would look back on their days together as a fine time that was often full of laughs. When he once prepared to spar with Marvis, Joe told him: "Rod, Marvis called your momma a bitch." Marvis grinned and said, "Awww, Pop, I never said that." On the road together, Joe would sit up talking with them in his hotel room, where he would sleep with the light on. Rodney remembered that they talked about "a lot of off-the-wall things."

"Rod!" Joe once said.

"Yeah, Uncle Billy?"

"Boy, if I die, don't let 'em take me to South Carolina and bury me there. Don't let 'em put me in a hole in the woods."

On another occasion, years later, they got into a conversation over a photograph Joe had seen in *Jet* magazine. "It was a picture of Olympic track champion Jesse Owens in his coffin," Rodney remembered. "And Uncle Billy said, 'Rod, if I die before you do, never let anybody take a picture of me in my coffin and put it in *Jet* magazine." Rodney shrugged and replied, "Okay, Uncle Billy."

Immodestly, Joe told his fighters they were blessed to have him in their corner, given his bona fides as a former champion. But even in an era of alphabetized "cheese champs," during which there was an expansion in weight classes and champions were crowned in the WBA, WBC, IBF, WBO, WBU, WBF, IBO, IBC, and IBU, Frazier did not develop a single champion in the 1980s and '90s. When it came to passing down his know-how, he only knew the way he had done it. He was not adaptable to fighters' individual skill sets in the way top trainers such as Durham, Futch, and Benton were. When he worked with young fighters, it was as if he saw within them the potential to re-create himself. The problem was, there was only one Joe Frazier. Few fighters have come along who have had "the whiskers" to battle in the trenches as Joe did and "breathe" on his opponents. Of the fighters Joe trained, Smokin' Bert Cooper came the closest to replicating him in build and style. For a heavyweight, Cooper was an undersized hustler with the same big heart Joe had inside the ring but none of the same discipline outside of it, where he battled a drug problem that undermined his career. Like Cooper—albeit taller and leaner—Marvis was on the smallish side for a heavyweight and would have been better suited to campaign as a cruiserweight. But Marvis was an exceptional athlete, and Benton had equipped him with the technical skills to take care of himself in the ring.

George James saw Joe work with Marvis at the gym. The Philadelphia trainer, who had once been stopped by the cops with Sonny Liston and had worked as a caretaker to Gypsy Joe, questioned Frazier on the approach he was taking with Marvis. "Why you got him down there crouching?" James asked. Frazier replied, "So he can get in there and roughhouse." James told me years later, "Marvis was over six feet. He should have been standing up and jabbing. But Joe had him fighting like a short guy—the way he used to fight. Like he was trying to put his own body inside of Marvis." According to James, the sparring sessions between Joe and Marvis were "unreal." Joe would tell him, "Get your hands up!" When Marvis took too long to get them up, Joe would slap him with an open glove. By undoing the work of Benton and rebuilding his son into his own image, Joe foreclosed on the promise Marvis had shown as an amateur. Larry Holmes told me, "Joe knew how to fight. His way."

No sooner had he gotten his career under way than Marvis found himself in a hospital gown. As he was preparing for his third bout in November 1980, he experienced a recurrence of the same issue that had been so alarming when James Broad stopped him at the Olympic trials in Atlanta. When Canadian heavyweight Gaston Berube landed a blow to his forehead and snapped his head back during a sparring session at the gym, Marvis spiraled to the canvas and remained there, conscious yet unable to move his limbs. As he had said in Atlanta, he was overcome with the same sensation that accompanies being hit in the funny bone, except that it spread across the length of his body. Surgery would be performed to correct the problem, which was genetic in origin. Doctors opened his neck and pulled a nerve away from his spine. With a six-inch scar descending from his hairline down his back, Marvis came back the following April. Among the eight victories he clicked off, he avenged his amateur loss to Broad and beat Joe Bugner in ten-round unanimous decisions. The thirty-three-year-old Bugner told the *New York Times*, "With time I think Marvis will become a very good fighter."

For a young man with just ten pro bouts behind him, Marvis had become a hot property. Along with the fact that he was the son of universally beloved Joe Frazier, Marvis possessed an array of attractive qualities. He was handsome, well spoken, and performed acts of charity in the community. He was a deacon at his church and visited prisons to conduct Bible studies. He even played Santa Claus at the gym for the children in the neighborhood. Though he had fallen short of his Olympic dream, he commanded early paydays that were commensurate with those of 1976 gold medalist Sugar Ray Leonard. As it once had for his father, "the love" poured in. The networks courted him. CBS bankrolled the Broad and Bugner bouts, paying him $150,000 and $200,000 respectively. Under the contractual agreement he signed with his father, he took home half of that, less taxes; Joe paid all expenses. According to Sharon Hatch, who oversaw contract negotiations for Joe Frazier, Inc., Marvis received a weekly draw similar to the one his father had received from Cloverlay, beginning at a hundred dollars a week and escalating as his purses increased. With the offers pouring in, Joe accepted one from NBC that would prove to be calamitous.

Overreaching with a young fighter can be perilous. Durham and Futch were assiduous in their choice of opponents, placing Joe in bouts where he could incrementally improve without undue concern about an unexpected setback. But when NBC dangled $1.2 million before his eyes for Marvis to face heavyweight champion Larry Holmes, Joe snapped it up. For Marvis, it would be like jumping from junior high school into a Ph.D. program. Undefeated in forty-four bouts, with thirty-one knockouts, Holmes was big, seasoned, and possessed a laser left jab. Even if one spotted some deterioration in his skills—as Joe did—or pointed to the fact that Leon Spinks had only had seven bouts when he beat Ali for the title—as Joe also did—it would not be under any circumstances the "picnic" Joe predicted it would be. Someone in the press corps looked up who Joe had faced in his eleventh bout. There it was, in the *Ring Record Book and Boxing Encyclopedia:* Billy Daniels, the barber from Brooklyn. Joe stopped him in six. Someone else asked Holmes if he had been prepared by his eleventh bout to take on the champion. He replied, "Are you kidding?" Now working with Holmes, Eddie Futch decried the pairing as premature by a year. "Why now?" he said. "Why risk a bad beating, the kind that could destroy him as a fighter?"

For Holmes, it was about the money. Even with no championship on the line—the WBC would not sanction it, given that Marvis was still unranked—Holmes was offered $3.5 million in NBC dollars. While he was happy to have the payday, he found himself in the same position he had been in with Ali: trying to avoid irreparably damaging an opponent who was clearly outclassed. Like Futch, Holmes had known Marvis since he was a young boy. Holmes had worked for Joe as a sparring partner for $350 a week prior to his second bout with Ali. Joe cracked his ribs. When asked years later what he had learned from those sessions, he replied: "Not to get hit. He would get in there to knock you out. He used to call me 'The Road Runner.'" As Holmes prepared to face Marvis at Caesars Palace in November 1983, Futch told him that young Frazier was like a son to him and that he did not want to see him get hurt, even as Joe himself still harbored hard feelings over Manila and the Bobick fiasco. Futch said, "Get him out of there as quickly as you can, Larry." Holmes liked Marvis, yet that fondness could and would not eclipse the fact when they stepped into the ring together, it would be "his ass or mine."

The fight was the brutal, one-sided affair that almost all had feared it would be. Stripping out of his green and gold robe with the words IN GOD WE TRUST on the back—the same colors that Joe had worn in the Fight of the Century—Marvis appeared dwarfed by Holmes as they stood eye-to-eye in the pre-fight instructions, Joe looking on ominously from over his shoulder. Nineteen pounds heavier and two inches taller—the very picture of a man in the prime of his career—the thirty-four-year-old Holmes answered the bell with grim focus. Tentatively, Marvis held his hands up high as he poked for an opening, then lowered them in a foolish display of showmanship, his chin exposed as he wiggled his shoulders. Holmes hurled him into the ropes. With his hands up again, Marvis maintained a perimeter as Holmes pecked at him with left jabs. When Marvis lowered his hands again, Holmes swatted him with a long right hand that spilled him to the canvas. Up at the count of eight, Marvis covered up and backed up into the ropes, where Holmes pummeled him with twelve unanswered right hands. He told me years later, "They were open-handed. They were slaps." With each one he threw, he glanced back at referee Mills Lane, at one point yelling: "Come on, man. Stop the damn fight." Lane finally did, at 2:57 of the first round.

Chagrined by the episode in which he had just participated, Holmes looked on glumly as Joe wrapped Marvis in a consoling hug. In an interview in the ring with NBC boxing analyst Dr. Ferdie Pacheco—who doubled as the matchmaker for the network and had ordered up this execution—Holmes said Marvis had not been ready and that he was glad he had not hurt him. While still in the ring, he stripped out of his trunks and robe and tossed them into the crowd. Later, with a sigh, he said, "This is a hard, dirty, rotten game." Appalled by the exhibition they had just witnessed, the reporters on hand pilloried Joe, who fielded their questions in the press room by firing back hand grenades. The "tyrant" Rodney spoke of came out. When someone asked if it had "hurt" him to see his son absorb the beating he had, Frazier replied that he had not been the one taking the punches. When someone else asked if he had "made a mistake" by sending Marvis in there with Holmes, he replied, "I never make mistakes." Overhearing that, Futch told Randy Galloway of the *Dallas Morning News:* "You'll never be able to convince Joe he was wrong. And anyone who disagrees with him on anything becomes the enemy." Uninjured, Marvis said in the dressing room again and again, "There is always tomorrow."

Marvis would work his way back up in the two and a half years that followed, only to find himself staring across the ring at a young and unbeaten Mike Tyson in Glens Falls, New York, in July 1986. Although Tyson had plowed through twenty-four opponents, twenty-two by knockout, Joe remained unimpressed, telling reporters: "He ain't nothin' to write home to Momma about." All week leading up to the bout, Joe kept it up, yapping the way Ali used to yap, almost as if it were him and not his son who had the date with Tyson. As Joe launched Marvis from his corner at the bell, Tyson charged from his as if he had been set free from a cage. Unable to take advantage of his six-inch advantage in reach, Marvis froze in place as Tyson drove him to a neutral corner. There, Marvis bent down and Tyson strafed him with right uppercuts that toppled him to the floor. From beyond the ropes, Joe implored, "Get up, son. Be a man, son!" Referee Joe Cortez began counting but got as far as three before calling a halt to the bout. Marvis said later, "Maybe God's telling me something." After three more inconsequential bouts, he retired at age twenty-seven, with some money but not enough to last. He would become an ordained minister in 1994.

No one can say for certain how far Marvis would have gone if his father had not intervened and rushed him along. While Marvis had some boxing ability, he would never be the "cold-blooded killer" in the ring that Tyson was or Joe himself had been. Even Joe knew that, occasionally scolding Marvis on his stool between rounds: "Do I have to get in there and fight for you?" Sympathetically, Sharon Hatch would always wonder if Marvis had it in him to even attempt to eclipse the deeds of the man he so idolized. "Joe was his hero," said Hatch. "Marvis could not see himself passing him or even equaling him in the ring. Joe was the champ. And Marvis was fine with being something less than that." *Newark Star-Ledger* columnist Jerry Izenberg observed, "Joe was trying to live through his son. I mean, how many men do that?" To Larry Hazzard, the New Jersey Athletic Control Board commissioner, even bringing up the subject of what could have been is an act of "Monday morning quarterbacking." That said, he added: "Maybe he did rush him. Going after the money is fine, but I think he could have slowed it down and still have gotten the money at some point." Years later, Marvis himself would only say, "Awww, I loved Pop."

———

Nothing had prepared Joe for the success he found. When Florence gave birth to their oldest child in 1960, they were so young and so poor that Joe looked down at Marvis in his crib and wondered to himself, "How'm I gonna feed this little bugger?" He would later say, "Man, I was worried about paying the rent, paying the light bill, putting food on the table." Daughter Jacquelyn followed a year later, and there were the two children he had with Rosetta: Renae (1960) and Hector (1962). Three other daughters with Florence came along later: Weatta (1963), Jo-Netta (1968), and Natasha (1970)—and suddenly, it was as if he were a youth again and the cousins were piling on his back as he lugged them across the backyard. But by then "the love" was flowing into his lap like a jackpot of coins and showed no sign of letting up, wealth larger than he or Florence could have ever imagined. Cheerfully, Joe spread it around in a big way, indulging not just himself, in cars and clothes, but those dearest to his heart and whoever else came to him along the way with a sad story. Close friend and business associate Darren Prince told me: "He lived not just in the day but for the day."

Upon boarding the Dog in 1959 in search of a better deal in the North, it would have been only understandable that he thought he was leaving the South Carolina Lowcountry in a cloud of exhaust. But the culture of poverty he had grown up with there would remain with him as if it had stowed away in the bag he had packed with his Sunday church clothes. It gave him his drive to excel in the ring and his love for song, yet it left him with the inability to embrace life beyond the horizon of today. Long-range planning yielded to the urgency of now, the only reality that he could be sure of in a world that promised no tomorrow. Even as his mother Dolly carried him in her womb, he had come chillingly close to expiring there when Arthur Smith shot into the car she and Rubin were in and narrowly missed killing her. Routinely, black lives in the Deep South were shortened by hunger, disease, and bigotry. While he would come into vast sums of unexpected money, his childhood in the Lowcountry became the lens through which he viewed his evolving circumstances.

Even as he found himself in sophisticated circles in the years to come, befriended and indeed adored by all manner of celebrities worldwide, he remained that same country boy. "He was happy eating a can of beans," said Richard Slone, who once trained under Joe and later became an artist. Though Joe had street smarts, he had come from a place where, instead of placing his few valuables under lock and key at a bank, his daddy buried them in the hog pen out back. By his own admission, Joe was "not good with numbers" and was uneasy dealing with the challenges that big money entailed. Cloverlay had helped him to a point, but when it dissolved and he was out on his own, he adopted a more or less haphazard approach to his finances, handing over control to whoever came along and won his trust. Whenever he was queried on the subject of his investments and such, he would say, "My job is to fight. I have other people who take care of the money." Content to have a few thousand stuffed in his sock, money he used for gambling and to dole out as necessary, he was never one to sit down with a financial planner, go over his portfolio, and develop an investment strategy. Only one question concerned him: *Is there enough on hand to pay the bills?* Former business manager Burt Watson remembered that he and Marvis would hold a weekly meeting with Joe to go over what he owed. "Marvis would say, 'Coach, we gotta get Pop, sit him down and catch him up with this love," Watson said. "Joe liked to hear that there was two to three hundred thousand dollars in the account. When it got down to thirty or forty thousand, he would say, 'That all we got? It's time to go to work.'" To augment his pension dispersal, he did commercials, personal appearances, and whatever else he could do to cash in on his brand. When their meeting was over, Watson would remember that Joe would ask for a check for a thousand dollars, cash it at the bank, pocket six hundred dollars, and slip the other four hundred in his sock.

Chaos engulfed Joe in the 1980s and '90s. In 1983, eleven days after Holmes whipped Marvis in Las Vegas, the Internal Revenue Service had Frazier up on the witness stand in U.S. Tax Court in Philadelphia to explain $220,000 in deductible expenses on his plantation in South Carolina. While Frazier had been quoted in the press saying he had purchased the property as a place for his mother to live, his former attorney Bruce Wright claimed it was a working farm and used it as a tax shelter to help offset the approximately two million dollars in taxes Joe paid from 1971 to 1976. Although there was an earnest effort to develop it as a farm, the high level of clay in the soil and the fact that the land was 40 percent swamp impeded that effort. Nephew Rodney Frazier would say years later, "And it had alligators galore! I remember when we once dredged the pond, we opened up one of their dens and they came crawling out. You have never seen so many alligators. We had to hide the dogs." Additionally, seventy-five head of beef cattle Joe purchased in 1974 died from intestinal parasites. Altogether, the plantation yielded only nine thousand dollars in gross receipts during the five-year period in question. The plantation would become just one of the tax issues that ensnared Joe, who sold the property for a loss in January 1988.

With all but one of their children sixteen or older, Joe filed for divorce from Florence in May 1985. It was not unexpected, given how strained the relationship had been for years. Florence had told him, "You call yourself a Christian, all this running around you're doing? You better get it together. God is watching you, Billy." Though Joe had been a generous provider to her and their children, he had cultivated an existence apart from them that would begin in secrecy and end in painful disclosures. In the period from 1980 to 1984, he had three children outside of his marriage: sons Joseph Mahoney and Joseph Jordan Frazier (Joe Jr.)—with Joan Mahoney and Sharon Hatch, respectively—and son Brandon Cottom, by Janice Cottom, a year before he asked for a divorce. Seven years later, in 1991, he had a final child—Derek Dennis Frazier, by Sheri Gibson—bringing his total offspring to eleven. The divorce proceedings between Joe and Florence would drag on for an extraordinary twelve years, which spoke to the complexity of the estate and the hard feelings that had emerged from what daughter Weatta called "years of pain and disrespect."

Unable to re-create himself in the ring, it seemed almost as if Joe had opted instead to do so through his heirs. In the same way Tudor sovereigns practiced procreation in deference to the royal lineage, he was only too pleased that he had sons. "Only sons could carry on his name," said Weatta, who paused and added, "He was weird that way." But he had come out of a world in which sons were prized as breadwinners by men and women who favored big families. Joe himself had grown up in a house with nine siblings—six of them boys—and that did not count the seventeen others whom he claimed in his autobiography that his father Rubin had sired. Had Joe not said years before that he was the son of his daddy? And that to change him, you would have to go back and change Rubin? Rodney said that in the eyes of his uncle, "having children determined what kind of a man you were." He remembered a conversation he had with Joe at the gym one day.

Oddly, it began with Joe calling out to him and asking: "Rod, are you gay?"

"Why would you ask that?" Rodney replied.

Joe eyed him and said, "Marvis has children. Mark has children. Hector has children. Why not you?"

"Because of you and my father," said Rodney.

"What the hell is that supposed to mean?" said Joe.

"My father had nineteen children by twelve or thirteen different women," said Rodney, who would later marry and have a daughter. "You had eleven kids by six different women. Every dime you get goes out the door to somebody."

Joe told him he was crazy.

"Paying for all those children?" Rodney said. "And not being with them every day to help them through the struggles of life? You call that being a man? I consider that being a fool."

"You callin' me a fool?" Joe snapped.

"No, Uncle Billy," he said. "Not a fool, just that you could have done things differently."

Weatta also chastised her father. Over his protestations, she told him in the 1990s: "Look, you have to stop having these babies and not marrying their mothers. You know what Renae and Hector have gone through? What kind of life they've had? Or Joseph Mahoney? Joe Jr.? Brandon? And now Derek?"

Sailing down the highway behind the wheel of one of his cars, Joe would roar the lyrics of a Johnnie Taylor song he was fond of that had personal significance: "The downfall of too many men is the upkeep of too many women." By the early 1990s, he had found himself steeped in the encumbrances of not just one woman but six, not just five children but eleven—and that did not include the attention he showered on Denise, who lingered in the background and adopted a strategy of letting him have his space with the hope that he would come around to her. While Joe endeavored to give more of himself to his children than just the envelopes he or his aides dropped to their mothers each month—he showed up when he was called upon for school plays, teacher meetings, and such—he found himself immersed in the unpredictable moods of women who not long before had found his charm so irresistible.

"Celebrity, power, and money are aphrodisiacs to women," said Sharon Hatch. "When you met Joe, you would think: I would love to be around him. Because he was so upbeat—he had that smile—and he would have you laughing." That said, Hatch understood that he would never be a "one-woman man."

"More was coming at him than he could have ever imagined," she said. "Look at it this way: You cannot take a child who gets one piece of candy a year, let him loose in a candy factory, and not expect him to eat all he can."

Even as "the love" slipped through his now gnarled hands, Joe did have a long-term investment into which Bruce Wright had guided him in 1973. With $843,000 from his boxing proceeds, Frazier purchased 140 acres in Bucks County that would later be developed into a housing development of 476 town houses called 100 Acre Woods. Frazier and Wright formed One Hundred Acres Ltd. According to Werner Fricker—whose Fricker Corp. purchased the acreage in 1978 and later developed it—Frazier had sold his interest in the land to A. M. Greenfield Trust, whereupon it passed into a series of other trusts. Under the terms of the sales agreement, Frazier was to be paid his share of the proceeds in installments over a twenty-year period. Until Wright died, at age sixty-nine, in April 1991, all went smoothly: Joe received his payments in an orderly fashion. But when Wright died, the 1991 payment never showed up. In an interview with *Philadelphia Daily News* reporter Leon Taylor, Jacquelyn Frazier-Lyde—an attorney who took up the case on behalf of her father and would later become a judge—said that "the chain of title was broken somewhere along the line as if, as far as that land was concerned, my father never existed." Frazier-Lyde added that she believed that annual payments her father had been receiving were not part of the land deal but that Wright was "just paying him from the proceeds he earned when he was fighting."

Lawsuits were filed and refiled. In November 1998, Frazier-Lyde sent a letter to the homeowners at the Northampton Township development that read, in part: "Please be advised that the premises on which you reside are subject to a claim of ownership by our client, Joseph Frazier." The letter claimed that Frazier "demands payment for this land" and stated that "all transfers of title, subsequent to his purchase of this land, are based on fraudulent transfers without consideration/payment to him for his land." For someone who had been so beloved in the community, Joe suddenly found himself in the crosshairs of public scorn. Common Pleas Judge Edward G. Biester ruled in favor of the homeowners in 1999. Frazier would later say that his "fight" was not against the people but "against the corporate scamboogahs." To stir up support, he was joined by activist Dick Gregory in a three-day march from his gym to Bucks County in September 2003. Former Fricker Corp. attorney Edward J. Hayes told *Philadelphia Daily News* reporter Ramona Smith that the claim of ownership by Frazier was "asinine," stating that while Frazier and Wright *did* agree to sell the property for $1.8 million, a later deal between Wright and Greit Realty Trust Co. superseded that previous deal. Joe claimed he never signed off on that later deal and that his signature had to have been forged. A friend of both would say, "Even if it had been unintentional—and knowing Bruce I think that it was, that there was some slipup on his part in recording the mortgage—Joe got screwed."

Larry Holmes had joined Frazier on his quixotic march to Bucks County. He told reporters that Joe was a friend and that he would do whatever he could to help him. Though Holmes would say he was not clear on "the whole story," he told me years later that he did know this: "If you give somebody control of your shit, you'll lose your shit."

Strapped behind the wheel of the Corvette that Joe had given him, Hector pressed hard on the gas as he barreled down the Atlantic City Expressway. In the passenger seat was Kevin Dublin, whom Hector had befriended when he spotted him in a street fight and told him, "You should stop by the gym sometime." Weaving in and out of traffic, Hector came upon a Mercury that would not let him pass. When Hector sped up, the other car sped up. Hector gunned it and got around the car, only to have it zip ahead of him. Dublin would say that it became "a game of chicken" at 110 mph, with both cars speeding up, then falling back, until the other driver veered in and Hector slammed on the brake. The Corvette soared eight feet in the air, flipped over on its T-top, and slid down an embankment onto the grassy median. Incredibly, neither Hector nor Dublin was injured. Even more incredibly, they were still able to drive the car on to Atlantic City, where Rodney had a bout that evening against James Broad at Trump Casino Hotel.

Joe looked at Hector and Dublin in disbelief when they walked into the dressing room. From head to toe, they were covered in grime. "What the fuck?" Joe said. "Where was you at? You come in looking like that?"

Dublin said they had an accident.

"What you do? Run into a pile of dirt?"

Rodney turned to Dublin and asked, "What happened?"

"Hector flipped the Corvette," said Dublin. "Almost killed us."

"Didn't I tell you?" Rodney said. "You keep fucking with Hector, you'll end up dead."

Even more than Marvis or any of his other sons, Joe saw himself in Hector. Before Joe found his way to Philadelphia and into boxing at the Twenty-Third PAL, he had been on the loose in New York. Hector stole cars and sold off the parts, just as Joe had. Rodney would remember how Hector had told him that when he was in

New York, he used to steal the catalytic converters from cars and sell them to junkyards. He could earn upwards of five hundred dollars per night doing that. "One night he got caught in the act," said Rodney. "The owner of the car came down, shoved a gun in his face, and told him to either tighten up every bolt he had loosened or he would blow his head off. He did it and the guy let him go." Only eighteen when he joined his father in Philadelphia, Hector even then possessed what Rodney referred to as "a jailhouse mentality." Rodney would say, "There is no other way to say it: Hector aspired to be a gangster."

Joe could not help but feel he had let Hector down. There had been an inequity between the advantages that had been afforded the five children he had with Florence and the two he had with Rosetta: the former had grown up in a big house with a swimming pool in a posh suburb of Philadelphia; the latter were raised in public housing in New York. Stitched into the robe Joe had worn for the Fight of the Century were the names of his children with Florence—and only their names. Renae and Hector looked upon them with some derision as "The Fabulous Five." As the two went through school, it had been awkward for them when someone asked them who their father was. When they would reply "Joe Frazier," no one believed them until they heard the backstory. They would say: *Say again, your father is who? Joe Frazier, the heavyweight champion of the world? Then why are you so poor?* Given the hard feelings Rosetta had toward him, Joe remained on the periphery of their lives until word got back to him of the trouble Hector was having. Marvis drove to New York and picked him up. "I think he had either gotten out of jail or was on his way to jail," said Rodney. "Uncle Billy said he owed him a life that was better than the one he was living."

Joe was old school when it came to parenting. Step out of line and he would threaten to send you back to Jesus. When it came to handing out discipline, he had more Dolly in him than Rubin. Like

Dolly, he was not one to spare the rod. In an interview with the Scripps Howard News Service in 1993, he explained his views on parenting, saying, "What I know mostly is that we need to put our foot down a little more, at home and everywhere else. Kids today need leadership." He even called on Washington to get involved: if a child was caught out after a certain hour, he believed that the parents should receive "a little summons" and a fine. As the years passed, he would become intolerant of the evolving fashions of urban youth. In laying down the law with his own children, he drew the line at cornrows, piercings, tattoos, and, as he told his youngest son, Derek, "wearing your pants down over your ass." Derek would remember challenging the prohibition on body art when he stopped into a tattoo parlor with some friends and found that he did not have the full fee. "So I called him and he rolled up in the Cadillac just as the tattoo artist was inking the last 'R' in 'Frazier,'" said Derek. "He was so furious at me that he could barely get out the words."

Tensions would emerge over the years with his children: Though they loved him—and he them—he had divided himself in a way in which none would get more than a piece of his attention. But when Joe was with them, he could be the same loving man that Rubin had been. He had that in him. Derek would speak of "the joy" his father had ushered into his life, and Joe would "drop everything" to be with him whenever he was called upon. When Derek was growing up, "Dad did not miss one event," said Derek. "Pick me up. Drop me off. Pick up my friends. Did everything." Observed Sheri Gibson, his mother: "Joe gave Derek more love than any man could give a child." Son Joseph Jordan Frazier (Joe Jr.) grew up in New Jersey with his mother, Sharon Hatch, and saw less of him yet said, "He would always call and check on me to see how I was doing. Or he would come over and see me. I still had him as a father figure, but just not as a daily presence." Hatch said, "Joe loved his children. As my son grew older, he and Joe spent more time together and got closer." But there would be quarrels, jealousies, and resentments among his offspring. Derek observed: "The problem was, there was only one of Joe, six women, and all kinds of kids. Everybody wanted to be the center of attention." Only four of his nine living children participated in interviews for this book, and only two of the five he had with Florence.

Conscious of the fissures, Joe had a jeweler carve up his Olympic gold medal into charms and gave them to his children. "He understood that not all of us were on the same page," said Joe Jr. "But his intention with the medal was that when he passed away, we would come together with our individual pieces and form a whole." Unmarried with a young daughter, Joe Jr. worked both in sales and with emotionally disturbed children. He supplemented his income by driving for Uber. "Sometimes when people hear who I am, they'll think I should have all this money and say: 'What are you doing driving for Uber?'" he told me. "On the other hand, people are very open to hear whatever ideas I have because of who I am." To unlock the door to his own potential, Joe Jr. immersed himself in personal-growth books and spotted aspects of his father in himself. Commitment and handling money were challenges for him, yet he would come to understand that his father and he were "links in a chain" that was forged across generations. "Unless someone comes along and breaks it, it just keeps growing longer," said Joe Jr. He paused and added, as if to clarify: "My father loved the only way he knew how."

Joe knew only what had worked for him when it came to rescuing Hector: get him off the streets and into the gym. Bringing Hector to Philadelphia, Joe also believed that Marvis would be a positive influence on him. Along with Rodney and Mark, Hector lived in the back of the gym, ran in the morning, and honed his boxing skills in a workout later in the day. He turned pro in February 1983 after a brief amateur career. Joe Verne, the owner of a wholesale furniture chain who teamed up with Joe to promote "The Fighting Fraziers," held a press event at Fuller Wholesale Meats, at Glenwood and Front Streets. There, in a setting similar to the slaughterhouse where Joe once worked, Hector, Marvis, and Rodney donned white overalls and protective gear as photographers snapped shots of them playfully swatting at the slabs of beef that hung from hooks. Fighting as a junior welterweight under the name "Joe Frazier Jr.," Hector beat a more-or-less average slate of opponents but would be stopped in his big test by future champion Vinny Pazienza in a nationally televised bout. He finished with a career record of 23-7-4 (nineteen KOs). "Hector was a helluva fighter," said Burt Watson. "He looked like Joe. He talked like Joe. He moved like Joe. He was Joe."

But the North Philadelphia streets called to Hector in the same way they had Gypsy Joe. At night, he would grow bored and sneak out of the gym in search of a thrill. Along with the Corvette he flipped, he would tear up the engine of every car that his father handed over to him. Like Joe, Hector possessed a wild craving for speed and loved to race. Rodney would remember he had a friend with a 1969 GTO Judge. "Whoever got out first usually won," said Rodney. "He and Hector would race all day long. Even before he flipped it, Hector beat that Corvette to death." Concerned that he was hanging with the wrong crowd, Joe told him, "Hector, if you don't stay away from those people, you're gonna write a check that your ass ain't gonna be able to cash." On that cold winter day in December when Joe stopped on North Broad Street and picked up the legless man, he meant to give Hector and Kevin Dublin a glimpse at authentic manhood. "Look at that man, out getting kerosene to heat his house and keep his family warm," Joe told them. "And look at you two scamboogahs. You got your arms and legs. Do something with them." What was Joe thinking as that tear slid from the corner of his eye? Of his own father and the courage he had shown in face of his disability? Or was it that even then a part of him knew that Hector was too far gone?

Some years would elapse before Dublin would come to appreciate the lessons that Joe had imparted to him. But when an eye injury ended his career, he began dealing drugs. "It was trifling and lazy," he told me. "But at the time I looked at it as a way of supplementing my income." Joe knew what Dublin had gotten himself into. Suddenly, Dublin was wearing an expensive fur and a watch, and was driving around in a Cadillac. "Where you get that watch?" Joe asked him. Dublin replied, "My girlfriend gave it to me." Joe eyed him and said, "Girlfriend. Okay, girlfriend. Remember son, there is the watch and the watchman." Joe had always been square with him. When he had gotten a girlfriend pregnant early in his brief career and was thinking of quitting, Joe had told him, "Let me worry about the Pampers. You get your ass back in the gym." Joe had given him a small weekly draw. But whatever friendship the two had was challenged when Hector once dropped out of sight and Joe came around to the corner looking for him.

"Have you seen Hector?" asked Joe.

"No, Smoke. Not for a while," replied Dublin.

"I know you know where he is," said Joe, his eyes blazing. "Look, if I got to shut down this corner, I will. This is about family."

Coolly, Dublin replied, "You do that and then it becomes a problem. Not just between you and me, Smoke, but for other people."

"I found Hector and sent him back to the gym," said Dublin, who ended up serving four years in prison. "Joe came by later and thanked me."

Hector tumbled into drugs himself. Over an eight-month period in 1989, he was arrested seven times for small-time burglaries. Burt Watson would remember that Hector once called the gym while in lockup and spoke to his father on speakerphone. "He talked of how tired he was of other inmates pushing him to the wall and fighting because of who his father was," said Watson. "Joe was stern with him. He told him, 'Nobody put you there but yourself. You got to come back to the gym.'" Weatta remembered that when Hector did show up later, her father vented his fury upon him. "What are you doing with your life?" he raged, as Hector stood chastened. "You've got the world by the balls and you're shitting on it. This is not who we are. We are not criminals. We're better than this. We're good people." But more trouble followed. On probation for a previous burglary, Hector was picked up by the police in January 1990 for stealing a bicycle from a ten-year-old boy and robbing him of three dollars. Hector told the court that he had a serious drug problem. He was sentenced to two eleven-and-a-half- to twenty-three-month prison terms plus five years of probation. When Common Pleas Judge Anthony J. DeFino asked why Hector had not been accompanied to court that day by his father, Marvis replied, "Our father is a hard man."

The sad tale of Hector Frazier would end years later in Sing Sing Correctional Facility in Ossining, New York. Upon leaving Philadelphia in the 1990s, he headed back to New York, got caught up in a car-theft ring, and was sentenced as a "three-time offender" to twenty-five years. Weatta remembered that her father refused to visit him until 2010, when she finally told him, "This is your child. You were no saint. That could have been you in there. You used to steal cars." Weatta, Marvis, and their father were joined by Renae at the penitentiary, where prison personnel and others spotted him and said, "Hey, Champ!" "My father was in pain," said Weatta. "On the drive up, he talked about how hard it would be to see his child locked up in a cage like an animal." Uneasily, they greeted each other in a public room, where tensions immediately boiled over. Joe told Hector he had no one to blame but himself for his predicament. Hector pushed back and asked him why he had not come to see him before now. But Weatta settled them both down and they spent two hours or so exchanging small talk. They then got some food from the vending machine and talked some more. When it was time to go, Joe hugged Hector and said, "Hold your head up, and keep the faith." Six years later, Hector died in prison of a suspected heart attack.

CHAPTER TWELVE

MAN TO MAN

Joe and Muhammad Ali at the 2002 NBA All-Star Game. *Philadelphia Daily News*, staff photo by George Reynolds

Curbside at Tulsa International Airport in April 1993, the limousine driver suddenly found himself in a predicament: Joe Frazier and Muhammad Ali's flights had unexpectedly arrived at the same time, and there was only one car to service them both. Ordinarily, that would not have been a problem. They could have hopped in the back together and he would have been on his way. But the driver had been told that under no circumstances was he to allow Frazier and Ali in the same enclosed space. Given the tense history between the two, the organizers of the charitable event that they had been flown in for were not taking any chances of a blowup.

Concerned that he could have Ali-Frazier IV on his hands, the driver presented his dilemma to Richard Slone, the young heavyweight from England who had come along with Frazier on the trip. Slone told Frazier, "Joe, the driver is in a sweat over there. Ali just landed. Can he jump in the same car with us?" Joe shrugged in approval. But when Ali had not shown up a half hour later, Frazier began to get irritable. "Where is he?" he said. He sent the driver into the terminal to look for him. Inside, the driver found Ali swarmed by fans as he handed out pamphlets on Islam. Apparently, he had come in by himself and had no luggage other than the briefcase he held in his hand. When Ali finally got to the car, according to Slone, Joe was so angry that he only nodded when Ali greeted him with a smile and said, "Hey, Champ." As Ali and Slone began chatting, Frazier turned up the volume on the cassette that he had slid into the car sound system, the rear of the limousine now filling with "Gina" by Bobby Womack. Joe began singing along.

"What are you doing, white boy, hanging around with Joe?" said Ali, leaning over to Slone so he could be heard.

Gina, only if I could turn back the hands of time . . .

"Joe has taken me under his wing," said Slone, then nineteen. "I'm one of his younger fighters."

"You run every day?"

"Three miles."

I would erase some of your doubts and fears . . . .

"You wear work boots when you run?

"I do, yeah. Just like Joe says."

"Good," said Ali. "Whatever Joe says, you do it. He is a great man."

That were put there by me, put there by me, my dear. Ahhh,
Ginaaa . . .

Hearing Ali speak of his affection for Joe warmed Slone to him.
He told Frazier so when they got to his suite. "He doesn't seem like
that bad of guy," Slone said. Joe shot him a look and asked, "What
do you mean?" When Slone told Joe what Ali had said about him
and that he had seemed genuinely respectful of him, Frazier told
his young traveling companion that Ali had "bullshitted everybody" for years and now "you fell for it, too." Instead of speaking
with Joe himself, Ali had always sent some version of an apology
through a third party. Even when Marvis stopped by his dressing
room in the aftermath of the Manila bout, Ali told him, "Tell your
daddy he is a great man." When Marvis relayed the message, Joe
told him, "Son, telling you is not telling me." So when Slone came
to him in Tulsa with yet another overture from Ali, Frazier was in
no mood to hear it.

"He did it with Marvis, and now he has done it with you," he
told Slone. "But man to man, he cannot come to me. If he had come
to me and said what he told you, I could respect him. But he has
never, ever done that. And he has had a million chances."

Just as their three fights had been a triptych for the ages, the
grudge that Joe held toward Ali in the years that followed seemed
to have the half-life of uranium. Being mocked by Ali as an Uncle
Tom, a gorilla, ignorant, ugly, and more had "cut me up inside,"
Joe told me years later. Even when it came to autographs they
penned on the same item, Ali would always sign above Joe, as if
to remind him of his superiority. Ali always pushed him to the
edge of his forbearance. But with the exception of the scuffle with
Ali on the Cosell set, Joe suppressed his anger under a veneer of
civility, content to settle whatever personal issues they had inside
the ropes. "Joe hated what Ali had done to him," said *Newark Star-
Ledger* columnist Jerry Izenberg, who had a close relationship with

both. "He hated that he did not have the verbal skills to parry with him." Izenberg once conveyed to Ali how Joe had blamed him for the cruel teasing his children had endured at school. Izenberg told him, "Didn't you think Frazier would be furious when his kids came home from school and told him their classmates had called their father a gorilla? Didn't you think he'd have a beef?"

Izenberg reported back to Joe.

"You spoke to him?" said Frazier.

"Yeah, I spoke to him," said Izenberg. "Of course, I spoke to him."

"And he talked about me and 'the gorilla'?"

"Absolutely, he did."

"Oh, what did he say?"

"He said that if he hurt your kids, or if he hurt your feelings, he was just trying to sell tickets and make money. He said he is very sorry."

"He said that?"

"Absolutely, yes."

Frazier looked Izenberg squarely in the eye and replied, "Go tell him to take that apology and shove it up his ass."

Philadelphia Daily News columnist Tom Cushman found Frazier to be just as intractable. At a retirement party that was held for him at the gym after his second loss to Foreman, Cushman spotted Joe standing by himself at the ring apron as his guests clustered in groups and sipped cocktails. While Cushman had been assigned by his paper primarily to cover Ali, he had gotten along well with Frazier through the years. "We had some conversations, just the two of us, and I got a feel for him," Cushman told me. "He was very straightforward." Given the glad occasion, Cushman sidled up to Frazier and told him, "You and Ali should be proud of what the two of you accomplished, the way you elevated interest in your sport. Even though I know you would have preferred a better outcome in two of your three bouts with him, you were a part of history." Frazier did not say a single word. He just fixed a stare on Cushman that had him "wondering if I would end up going down for the count." Cushman added, "He wanted Ali on his back. That was it. Just having been a part of it was not acceptable to him.

"Joe just did not understand where Ali was coming from," said Cushman. "And Ali did not understand why Joe did not understand."

Given that that their purses had been guaranteed, Joe would say that the "garbage" Ali spewed prior to their bouts had been un-necessary. But Ali never saw an audience of more than one that did not bring out the performer in him, and he would not stop until he had worked it into a frenzy. On the day *New York* writer Nik Cohn showed up for his workout at Deer Lake before the Manila bout, Ali had the largely white crowd so in his thrall that when he called Joe "a gorilla," they chimed in and called him an "ape-man." Ali would always claim that he was only "putting asses in the seats," but did he understand the power behind his words, how he had un-caged something primal and bigoted in his fans? Cushman would remember those scenes at training camp. "There would be a crowd of thirty, forty, fifty, or perhaps a hundred people at his workout, and he would be up in the ring screaming," said Cushman. "And as soon as he was done with that part of the day, he would go back to his dressing room, lay down, and become an entirely different person." But Joe had only one face that he showed the world, and it was an uncomplicated one, even if the inner man was far from that. Guided by the Golden Rule—"Do unto others as you would have them do unto you"—Joe was not unforgiving and seldom held on to hard feelings, except when it came to Ali. The antipathy he harbored for Ali simmered just below a boil.

A wrong word on the wrong day could activate his hot button. While driving down to South Carolina with Joe in the 1990s, Burt Watson offhandedly commented, "Whatever they said about Mu-hammad, true or not, he was one heck of fighter." Joe did not reply and appeared to let it pass. At South of the Border, Joe stopped at a gas station and remained in the car while Watson used the lavatory. When Watson came out, he discovered that Joe had left. "I'm thinking, 'He's staying at the pumps to get some gas,'" said Watson. "He got the gas. He got the gas and took off." Since cell phones were not yet in common use, Watson had no way of calling him. So he just stood there, perplexed. Finally, Joe drove up. When Watson got in the car, it became clear to him what had happened: By complimenting Ali, Watson had so infuriated Frazier that he had to drive around to cool off. Joe looked over at him in the passenger seat and told him, "Coach, if you think it, or ever feel it, keep it to yourself. I never want to hear it again."

Early in his history with Ali, Joe had said, "Let's see who will wind up the better man in his older age. Let's see who will wind up with the larger piece of the cake." By the twenty-fifth anniversary of the Fight of the Century, in 1996, the years weighed on both of them. Well along in his battle with Parkinson's disease, Ali had become enfeebled, his speech slurred almost to the point of being unintelligible. On legs that had once carried him with the elegance of Nureyev, he now shuffled with the labored gait of a man in his eighties. With hands that had once blinded opponents in a blur of leather, his grasp was now so unsteady that he had trouble picking up a glass. With each birthday that passed, he had become farther removed from the Ali that the world once knew and appeared in even worse shape than Joe Louis, who as a young man Ali had sworn he would never end up like. As he traveled the globe on behalf of humanitarian causes, Ali became beloved in a way that had seemed inconceivable during his stand against the Vietnam War. But Joe had no pity for him and said so. In far better shape than Ali physically—even as his own speech became harder to understand—he assumed authorship of the condition Ali was in and pointed to it as evidence that he had come out on top. I remember hearing him say so one day at the gym.

"Him and me had three fights," he told me. "He won two of

them. I won one. But if you look at him now, you can see who won them all. Me!"

Rage spilled from him in *Smokin' Joe*, the autobiography that he penned with Phil Berger, the former *New York Times* boxing writer. Published in April 1996, it heaped thirty years of scorn upon "Clay" in a slim 213 pages. Frazier wrote: "I had fought my way up from nothing. I'd earned my way with hard work, work that was owed respect. But this scamboogah thought nothing of talking about me as if I was some head-scratching dumb nigger." Moving on: "Well, to hear it from Clay, I was a lame specimen of a black man, a kind of Stepin Fetchit in boxing trunks." Elsewhere: "It never ended with this chump. Nonstop bullshit. While the public found it amusing, I guess, and came to view him as a good guy, I knew different. This was a nasty, envious, mean-spirited egomaniac. . . ." On and on it flowed, a virtual Niagara of bile. "Truth is," he continued, "I'd like to rumble with that sucker again—beat him up piece by piece and mail him back to Jesus." He would have even more to say, and he would do so in July at the Olympic Games in Atlanta, where Joe had a young boxer, Terrance Cauthen, in the lightweight draw. (Cauthen won the bronze.)

An unexpected and deeply moving scene unfolded at the opening ceremonies. With eighty thousand spectators in attendance at Olympic Stadium and 3.5 million viewers looking from around the world, U.S. swimmer Janet Evans carried the Olympic torch up a steep incline until she came to a halt on a platform. There, a roar went up as Ali stepped from behind a wall, his hands shaking as Evans passed the flame to him. As Ali lit a wad of flammable material at the bottom of a pulley, which would carry the flame to the cauldron atop the stadium, NBC announcer Bob Costas observed from the broadcast booth: "Look at him. Still a great, great presence. Still exuding nobility and stature, and the response he evokes is part affection, part excitement, but especially respect." But Frazier did not feel any of that. That Ali had been chosen to light the Olympic cauldron was galling to him. And when *Philadelphia Inquirer* columnist Bill Lyon came to him for comment three days later, he said so. He growled: "If I'd been up there with him, I'd have pushed him in the fire."

Unseemly and beneath a man who himself had been beloved, it was a stunning comment that had people wondering if they had heard him correctly. But there would be more. "You think he represents the Olympics or America? He hates whites." He groused that Ali had been a draft dodger and again pointed to the disparity in their physical condition, noting: "They said I was the one who was beat up by him, but look at me compared to him. Look at us." Lyon said in his column, "I have known Joe Frazier for more than two dozen years and I was floored by his attitude toward Ali. I suggested, almost urged, that he reconsider what he had said. . . . But Frazier not only refused to recant, or reconsider or soften what he had said, he repeated it, and then added: 'Say it loud and proud, brother. Tell it straight, what I said, the way I feel.'" One of the letters that poured into the *Philadelphia Daily News* called Joe "a washed-up has-been" and "a bitter, jealous old man who wishes he could have gotten the respect Ali gets." Even Joe Jr. later told him, "Dad, you can't do that. That's not right."

Joe would apologize in an interview with the Associated Press in October. He explained, "A man calls you all kinds of names, what are you supposed to do, stand up and take it on the chin? No. I had to fight back." Of his ongoing feud with Ali, he said he hoped the two of them could bring it to an end, adding, "We got to do it, before we all close our eyes, because I want to see him in heaven." Talks were under way for their daughters Jacqui Frazier and Laila Ali to face one another in the ring, and he looked forward to seeing that happen. (Laila beat Jacqui in June 2001.) But Joe still harbored a grudge that only seemed to become inflamed when Ali apologized to him in the *New York Times* in March 2001. Close friend and promoter Butch Lewis advised Joe to bury his grievances with Ali, if only to create business opportunities for the two of them. Others urged Joe to let go of his anger at Ali because of the example that he would set: if these two embittered rivals could settle their differences—if that was possible—it would invite others to do the same. But Ali still had not come to him face-to-face, and it was an omission that preyed upon Joe. When agent Darren Prince told him that Ali and his wife, Lonnie, had invited him to walk the red carpet with them for the January 2002 premiere of the biopic *Ali*, Frazier blew up at him, shouting, "You tell that SOB that if he wants to make up and shake my hand, he has to come to my turf to do it!" Prince would say that he had never seen Joe so angry.

"We should do this, Pop," said Marvis, who always had a fondness for Ali. He had once said, "Deep down, I believe they both love each other."

Prince called back later to see if Joe had reconsidered. Marvis told him, "Prince, I think we better back off. Pop is real, real angry."

———

One year back in the 1990s, promoter Joe Verne arranged to bring Joe and former champions Michael Spinks and Joey Giardello to Children's Hospital of Philadelphia for a Christmastime visit. None of the youngsters were old enough to know who Joe was, but their parents did and were delighted to meet with him and chat. For as long as he had been in the public eye, Joe never said no when it came to giving of himself with hospital visits and other charitable causes. "He would have done it every day if he could," said Verne, who placed Joe on salary to do appearances for his chain of furniture stores because of his way with people. "Nobody had to ask him to go. He would go on his own."

Over a three-hour period at the hospital that day, Joe and his group stopped in rooms to sign autographs and pose for snapshots. When a doctor approached Verne and asked if he could take them to another floor of the hospital, Verne replied, "Sure. I've got to call somebody. Go ahead, and I'll catch up to you." Only when he finished his call and found the floor they were on did Verne realize the doctor had escorted them to the cancer ward. Verne looked around but saw no sign of Joe or the others. He found a nurse.

"Where are the guys?" Verne asked.

Gesturing to a door that led to a fire escape, the nurse replied, "In there."

"The fire escape? What are they doing in the fire escape?" Verne asked.

Verne walked over and opened the door.

"All three of them were in there crying like babies," said Verne. "Joe grabbed me and said, 'Please get me out of here. Whatever you need me to do—go to another part of the hospital, go have lunch with the parents—I will do it. But not this. Young kids should never have to go through this.'"

Even as he revealed the worst of himself in his angry entanglements with Ali, Frazier remained connected to his more virtuous self. Through the years, it was almost as if there were two Joes. One remained at the command of his ego, of which his daughter Weatta once said, "Being a boxer, you have to have an ego. You have to believe that no one is your equal." But the other, away from the ring, was driven not by his granite will to stand apart but by the belief that he was one of many. When he once spotted a motorist stranded along the highway and his nephew Rodney encouraged him to just drive on, Joe's reply seemed to sum up how he perceived himself in the world: "Whoever that is could be your father or your mother, your son or your daughter." And when he slipped that "love" out of his sock and into the hand of that legless man in the wheelchair or did the same with Gypsy Joe and countless others along the way, it animated an aphorism that he was fond of using: "Never look down on a man unless you are helping him up."

In a city that developed and cherished a fantasy crush on

the fictional boxer Rocky Balboa—the South Philadelphia white heavyweight who, portrayed by Sylvester Stallone, emerges from obscurity to claim honor, if not victory, in a bloody showdown with the audacious black champion Apollo Creed—Frazier inhabited the ethos of the underdog in ascent. Even certain scenes from the original film appeared to have been extracted from his life. When he was a young unknown, he also slammed his bare hands into the animal carcasses at the slaughterhouse where he worked, and it was not uncommon for him to the run up the steps of the Philadelphia Museum of Art. While it would sometimes seem that every boxer who ever walked into a gym saw themselves in Rocky—including Chuck Wepner, the journeyman white heavyweight who stood toe-to-toe with Ali in 1975 before losing by a knockout with nine seconds left in the fifteenth round—Frazier would say of Stallone, only half joking: "That scamboogah stole that stuff from me." Joe had a cameo in Rocky and auditioned for the part of Clubber Lang in Rocky III. Stallone would remember on Instagram that he had a sparring session with Frazier but soon discovered it was "very foolhardy, hazardous and homicidal."

Stallone continued: "Once in the ring, I figured I [would] just move around and avoid his punches. That idea worked well for about two seconds. Simply because the next thing I knew there was a thunderous left hook that was planted extremely deep in my body. And an overhand right that resembled a falling piano landing just below my left eye. The world was now spinning in several directions."

Stallone ended up with six stitches.

Mr. T ended up with the part.

And the city of Philadelphia would end up with the statue of Rocky that appeared in the picture. For years, there was a public debate over where it should be located. Given that it was a movie prop that was donated to the city by Stallone and not actual art per se, it was not looked upon by as worthy of a place atop the steps of the Philadelphia Museum of Art, which Rocky had scaled in triumph. The piece was moved around before it was ultimately placed in an area adjacent to the base of the steps, where visitors have come from across the world to run themselves and stand with the statue for a snapshot. Only after his death did Joe himself get the long-overdue statue that now stands in the South Philadelphia stadium complex.

Underdog Joe would have given a leg up to Rocky had he been an actual person. Whenever Frazier came upon someone who reminded him of himself, a guy in search of a break as he had once been, he found a way to fold him into his life. "Joe collected strays," said Burt Watson, one his business managers. "He befriended people along the way." Kevin Dublin, his former fighter, would remember that in the neighborhood surrounding the gym, Joe became "a kind of Pied Piper" as he roamed from stop to stop. On any given day, he could be spotted picking up his cleaning, going for chicken wings, or hanging out with the firemen at Engine 50, Ladder 12, where in the summer he would play half-ball in the street. At the corner tavern, he would occasionally drop by, throw down some bills, and instruct the bartender, "Pour everybody a drink!" Up on North Thirteenth Street, he would show up every day to play the "street number." In a booming voice, he would announce his play, which would always be the same: 11111 00000 5555. "He hit now and then," said Darren Renwrick, who booked his action. "Once for four thousand dollars. When he won, he always gave us a handsome tip." Unable to get there on a particular day, he sent around his driver, Legrant Presley.

Dublin cracked up in laughter as he told the story. "The number hit but when it came time to get paid off, Legrant was nowhere to be found. He disappeared. Smoke was furious. He said, 'I know that scamboogah took that money and spent it.' Finally—two weeks later, I guess—Legrant got the courage up to come around. He swore to Joe that he never got around to playing it. Legrant probably just figured that the number would never come in. . . . Smoke forgave him, but it took him a while to get over it."

One day in November 1989, Joe showed up at the "lottery house" Renwrick ran with talk show host David Letterman, who had come to North Philadelphia with a film crew to shoot a segment called "A Day in the Life of Joe Frazier." Earlier in the year, Letterman had cooked up a comic sketch that involved sending Joe a hundred dollars to fill up his tank with gas and then checking back with him during the course of the week to see how he used it up. "Every show, Letterman would call Joe and adjust the level on a cardboard cutout of a gas tank," said Watson. When Letterman came in from New York with his crew, Renwrick would remember that he had an L.A. Dodgers cap down over his eyes, waved a cigar, and wore a raincoat with an upturned collar, "just like Peter Falk in *Colombo*." Joe was dressed in a sharp blue blazer over a red sweater, with chains around his neck. Letterman accompanied him on his daily errands, which included buying gas, dropping off his laundry, and having lunch. Over his plate of food, Joe said a prayer that ended with an appeal to the Lord to "keep the scamboogahs" off their backs. Letterman paused a beat and observed, "That was beautiful, Joe." While Joe was no more of an actor than a singer, he handled the spot with charm.

People gravitated to Joe in the same way they did to Ali, perhaps not in the same swarms but with the same genuine affection. "Joe and I had a thousand dinners together, and I never saw him finish a meal," said Verne. "When fans came up to him in a restaurant, he would shove his plate aside, sign an autograph or stand with them for a picture, and ask them to sit down for a beer." When Verne once asked him why he was always so accommodating. Frazier told him, "Because these are the people who made me who I am." Wearing a cowboy hat, an array of jewelry, and a spritz of his favorite cologne, Aramis, he would stride into a casino or a club and immediately feel at home. Occasionally, someone would step up and ask him to arm wrestle. "He beat everybody," said Michael Averona, who remembered how he once challenged Frazier himself. Joe told him: "We don't arm wrestle. We're buddies. You don't want to beat your friend." But when he was so inclined, he could drink even his closest friends under the table.

The "game" was called Man or Mouse. "Say there were four people at table with four different drinks—gin, vodka, scotch, and brandy," said Watson. "Joe would pour each of them into an empty glass, stir it up, and take a big swig. Then he would go around the table and ask, 'Who's gonna be a man and who's gonna be a mouse?'" To which Watson once replied, "Boss, I'm feeling like a mouse today." Cardiologist Dr. Nicholas DePace remembered seeing Joe mix wine and bourbon together and drink it down. "He was drinking a lot when I met him in the 1980s," said DePace, who owns one of the largest sports memorabilia collections in the world and purchased an array of pieces from Joe. "Drinking to him was more of cultural exercise that dated back to his youth in the South, where they used to pass around the moonshine jug as a way of socializing. But he was not what I would describe as a classic alcoholic, the way you would think of someone who ends up on skid row." Denise agreed that he indulged in "crazy drinking" yet stopped short of characterizing it as "alcoholic drinking," explaining, "When he partied, he partied hard. He is the only person I ever knew who could drink a bottle of vodka, get up, and not even have to take an aspirin." Daughter Weatta said, "By the 1990s, he had cut down on his drinking dramatically. I think he just asked himself: 'Why am I doing this?'"

Friendships were sacrosanct to Joe. Gloria Hochman was once in the hospital when he dropped by to see her unannounced. "Stan had told him I had gone in for a procedure," said Gloria. "Joe showed up dressed in a long fur coat and sat on the edge of the bed. You should have seen the look on the face of the woman in the other bed." Nephew Mark Frazier remembered joining his uncle for a drive to New York to see his old friend Emile Griffith, who Joe had befriended when they were in training camp together prior to the opening of the new Madison Square Garden in 1968. "Emile was in an extended-care facility and could barely speak," said Mark. "But you could see in his eyes how happy he was that Uncle Billy dropped by." George Kalinsky, the Garden photographer, would remember Joe as fiercely loyal. Kalinsky said that when his daughter was married back in the 1980s, Joe showed up at the reception because years before he had promised her that he would sing at her wedding. Some ten years later, Joe showed up again when Kalinsky buried his wife. "On the day of the funeral, there was an incredible snowstorm in New York," said Kalinsky. "But Joe and Marvis got in the car and somehow got there in time for the service. Seeing them trudge into the church with snow on their shoes was extremely touching."

Fifty-one years after he departed on the Dog, Joe came back in September 2010 to receive the Order of the Palmetto, the highest honor that can be bestowed upon a civilian in South Carolina. At the ceremony, held at the Henry C. Chambers Waterfront Park, near where, years before, Dolly Frazier had peeled and deveined shrimp from the ocean trawlers, Governor Mark Sanford observed, "It is magnificent when you think about the odyssey, the journey that has been his life." Frazier had asked Kalinsky to come along with him on the trip, during which he visited with relatives and old boyhood friends. Along with all but two of her children—Joe and Mazie—Dolly had been gone now for ten years, led to her crypt deep in the woods by horse-drawn carriage. Twelve hundred people had come to her funeral. When Joe and Kalinsky stopped by the house where Joe was born and grew up, they found it weathered by age but still standing. In the back yard, the hog pen was gone but the big oak where Joe hung his do-it-yourself heavy bag filled with rags and corncobs was still there. Upon seeing it again, he walked up to it and wrapped it in a hug.

———

Finally, Denise had Joe to herself. Although it was not the big house on the hill that he had once promised her, the twentieth-floor apartment that she had found them at the Windsor Arms on the Ben Franklin Parkway was more than enough room for the two of them. For Joe, it was an overall better arrangement than the one he had at the gym, which now had a history of unpaid taxes and had become increasingly hazardous with each passing year. Richard Slone would remember an occasion when Frazier slipped and went head-over-heels down the steep, narrow steps that led to his living quarters. "When he got to the bottom, I thought he was dead," said Slone. "But he popped up and with a laugh said, 'They're tryin' to get me.'" Emotionally, leaving the gym would be a sad parting. But the Windsor Arms had an elevator, and he had Denise there to care for him, which she did in the same unstinting way she always had.

The years had flown by since Denise had met him at the City Squire in New York, after the Mathis fight. She had been so young then. When she looked back at herself, she would sometimes think how "ridiculous" it had been for her to follow her heart and not her head when it came to her relationship with Joe. For years, good friends would urge her to see other men, and occasionally, when she and Joe were on the outs, she would go on dates and pretend to enjoy herself. But she had fallen in love with Joe and would always remain in love with him, even though she found herself troubled as the years passed by an unintended consequence of their affair. "The time he spent with me was time he could have been spending with his children, taking them to the movies or whatever," said Denise. "Now that I am older, I can see how selfish that was of me." Weatta said that Denise apologized to her and that she appreciated that. "Look, it takes two to tango," said Weatta. "My father should have married Denise once he and my mother were divorced." But it was enough for Denise that it was just the two of them now, and that they would grow old together. Sadly, there would not be enough time.

Even well into his sixties, Joe had been spared the tragic outcomes of his profession. With the exception of George Foreman— who had struck gold pitching grills and mufflers—nearly all of his most celebrated opponents came to a sad ending. Eddie Machen: plagued by clinical depression and dead at age forty from a fall from a second-story apartment window. Oscar Bonavena: gunned down at age thirty-three outside of a brothel in Reno when the owner caught him sleeping with his wife. Buster Mathis: beset by an array of health problems as his weight ballooned to 550 pounds, the victim of a heart attack at age fifty-one. Jerry Quarry: unable to dress or feed himself, claimed by dementia pugilistica at age fifty-three; his younger brother Mike would die at fifty-five of the same syndrome seven years later. Jimmy Ellis: he also battled dementia pugilistica and would outlive Joe by three years, dying at age seventy-three in 2013. And Ali: he appeared to become only more diminished as the years went by. By comparison, despite having cut off a big toe while mowing the lawn and having severely burned his arm taking the cap off a still-hot car radiator, Joe seemed to be doing better than he could have hoped for. Even so, he battled diabetes and hypertension and was in and out of the hospital for surgeries on his neck, shoulder, and back. Lens implant surgeries were performed by Dr. Yanoff in 1997 and 1998 that corrected the vision in both of his eyes to 20/20.

Doctors who examined Joe in his later years at the Hospital of the University of Pennsylvania observed that he had developed "significant cognitive impairment." Given the warrior style that he favored and the punishment he had received, it is likely that he also came away from the sport with some degree of chronic traumatic encephalopathy (CTE), a progressive degenerative disease found in people who have suffered recurring brain trauma. Common among boxers who have had long careers, it was known as dementia pugilistica since it was addressed by Dr. Harrison Martland in the *Journal of the American Medical Association* in 1928. By virtue of the work performed at the CTE Center at the Boston University School of Medicine, in conjunction with the Concussion Legacy Foundation, it has since been discovered in football players and athletes in other contact sports. Of 111 former National Football League players whose brains were autopsied, 110 of them were found to have CTE, which can only be detected after death by an autopsy of the brain tissue. It has been linked to an array of behavioral symptoms that could be found in Frazier, including a lack of judgment, explosive anger, anxiety and depression, binge drinking, and sexual promiscuity. According to Dr. Robert Cantu, clinical professor of neurology and neurosurgery and cofounder of the CTE Center at the Boston University School of Medicine, the presence of these indicators would suggest "an overwhelming likelihood of CTE." Dr. Cantu added, "I would say it is consistent with the profile we so often see, but we would not be able to nail that down unless an autopsy was performed." Frazier would not receive an autopsy.

Joe was in a good mood when I visited with him at his apartment in June 2009. He was sixty-five and still on the go. As he sat at his dining room table and picked at a bowl of cherries, he told me he was leaving in a few days for the United Kingdom for a series of appearances. It was a trip he had been both looking forward to and dreading. Flying still spooked Joe. Were it not for the small matter of the Atlantic Ocean, he would have packed up and driven to the UK, even if it required extending his journey by days and days. Sheepishly, he said, "If there was only a way I could just snap my fingers and be there." On the wall above a sofa hung the classic photograph from the Fight of the Century of Ali buckling to the canvas in the fifteenth round. Joe looked over at it, chuckled, and said, "There he goes!" Thirty-eight years had passed since he had delivered that blow and still it gave him a thrill. I asked if it upset him that Ali aide Dr. Ferdie Pacheco had called him "dumb" in the documentary *Thrilla in Manila*. "Ah, they rip me so Muhammad can come out smelling like a rose," he said. "Guys who were on his payroll, they would say anything. And Muhammad, we were friends back in the beginning. When he was out of boxing for dodging the draft, I went down to see President Nixon to help him get his license back. I went along with him because he seemed sincere. But whenever a crowd was around, he would go off on me and shout: 'Joe Frazer is no champion!'"

Joe chose a cherry from the bowl and sucked on it. "He said once I would have been nothing without him. But what would he have been without me?"

Less than two years later, Joe's health began to fail. When he flew to Las Vegas in September 2011 for the Floyd Mayweather–Victor Ortiz bout at the MGM Grand, he appeared gaunt as he sat at a promotional event with Ken Norton and Leon Spinks and signed autographs. Richard Slone, who now lived in Las Vegas, had not seen him in five years and had prepared an itinerary for him that would have been in keeping with the pace he once set: they would have dinner, hit a casino, and perhaps stop at one of the local strip clubs. But as soon as he saw Joe, Slone could see that he was not up to the chase. When Slone picked him up at the South Point Hotel and Casino, Joe told him, "I think somebody poisoned me. Everything I taste is like metal." At an Asian restaurant that evening, Joe ordered as lavishly as he always did yet could only eat a half cup of soup. Eyeing the unfinished food, he told Slone, "Get them to pack it up." Slone would later say, "Seeing him in that state was very shocking." At McCarran International Airport that Sunday, as Joe sat in a wheelchair and waited for his flight back to Philadelphia to be announced, L.A. matchmaker Don Chargin spotted him and stopped by to say hello. Forty-five years had elapsed since Chargin had worked with him on his bout against Eddie Machen.

"Whenever he saw me through the years, he called me 'Machen,'" said Chargin. "When I saw him at the airport, he was in a wheelchair and told me that he thought he was coming down with something. He said, 'Hey, Machen, go get me a bottle of water, would ya?'"

Within days of arriving back in Philadelphia, he was seen at the Hospital of the University of Pennsylvania. Denise was by his side. By now, he had shed thirty-seven pounds in less than a year. Initially, it was suspected that his weight loss and poor appetite were related to his "relatively mild" dementia, yet in the assessment of his attending physician, a "more sinister condition such as [a] . . . malignancy" could not be excluded. On September 23, he was diagnosed with hepatocellular carcinoma—liver cancer—which develops in a setting of chronic liver inflammation and is linked to chronic viral hepatitis infection or exposure to alcohol. The tumor was large and inoperable. In the ensuing weeks, he was in and out of the hospital, where he underwent chemoembolization to stem the blood flow to the growth and later palliative radiation therapy to keep him comfortable. With the help of her sister Trudy, Denise saw to his needs as his condition deteriorated in the final weeks. Trudy would remember that while he was still able to speak, she and Denise one day joined him in song. "We were in the room with him and he turned on his boom box," said Trudy. "We held hands with him and sang 'My Way.'"

Even loved ones were discouraged from stopping by and seeing him. "Joe did not want anyone to see the condition he was in," said Denise. But he asked Rodney to get him a bottle of Aramis, his favorite cologne, and Michael Averona stopped by the apartment with a prayer blanket. Averona would remember, "He was a shadow of himself. Sickly, withdrawn. I looked at him and started to cry. He pulled the covers back, scooted himself up, and raised his left fist in the air. He said, 'Don't do that. I'm gonna be all right.'" But he soon slipped into stillness. Marvis sat with him and held his hand in prayer. With only hours remaining, Weatta and her husband Gary Collins came by with Mazie, his only remaining sibling. She had been with him sixty-seven years before when he had come into the world on that January day in Beaufort and she was with him at the end. On the evening of November 7, 2011, he passed away with Denise lying at his side.

"Daughter, be sure they clean me up and dress me nice," Joe had once told Weatta. Denise chose a blue suit for him to be buried in. She asked Rodney—who cut hair in his spare time—to give his uncle a trim at the funeral parlor. Sensing that he was apprehensive, the undertaker told him, "Rodney, the only way you are going to be able to do this is if you talk to him like he is still alive." So Rodney secured a cape over his chest and began conversing with him. "Uncle Billy, I'm going to shape you. I'm going to clip the top of your head and then your sides." Rodney then groomed his mustache and got out the bottle of cologne that he had purchased for him. "Uncle Billy, you asked me to get you some Aramis," Rodney said. "I never got a chance to give it to you, but here it is." He sprayed him with it, packed up his gear, and left.

Four thousand people attended the funeral at the Enon Tabernacle Baptist Church. Helping to defray the cost of it was Floyd Mayweather, who volunteered close to seventy thousand dollars. Florence and her children came to the church in one car, Denise and Trudy came in another, and Sharon Hatch, Sheri Gibson, and Janice Cottom and their children came in still another. Ali—five years before his own death in June 2016—was guided to a seat by his wife, Lonnie, as other stars of the boxing world looked on: Larry Holmes, Leon and Michael Spinks, Bernard Hopkins, Gerry Cooney, and others. Civil rights activist Reverend Jesse Jackson called Frazier "an ordinary Joe who did extraordinary things." When Jackson asked the attendees to "rise and show your love" in remarks during the service, Ali stood unsteadily and clapped. At the end of the three-hour service, the remains were taken to Ivy Hill

Cemetery in Philadelphia and placed in a temporary crypt. Later, they would be accorded a memorial more befitting of a champion: a handsome granite crypt with a picture of Joe Frazier in his boxing stance. Surrounded by flowers, it is just what Joe would have hoped for, a place far from South Carolina and its deep, dark woods.

———

Someone once said that to bear a grudge and not forgive someone is like swallowing a cup of poison and expecting the other person to die. I occasionally thought of that phrase as I pondered the life of Joe Frazier, a good man but not a perfect man, and wondered how long he held on to the anger he carried. Did he ever find a way to let it go and forgive, or did it just eat him up? I was hopeful that it would be the former and came upon a heartening clue when Eva Futch sent me a copy of a letter that Joe sent her husband, Eddie, in 2000. The occasion was an award Eddie was to receive from the NAACP. (Futch passed away the following year, at age ninety.)

The letter read:

> *Eddie:*
>
> *Please forgive me for not being able to be with you on this special occasion. Receiving an award from the NAACP is a great honor.*
>
> *You are truly a man worth honoring. You are loyal to the titles father, husband, friend and Trainer. Ever since I have known you, you have been nothing less than compassionate. And in the sport of boxing, that is something you rarely see.*
>
> *I remember the 'Thrilla in Manila,' the fans were on their feet and the butterfly and I were at war. We had both taken severe punishment and I could not see out of my eyes. Between the 14th and 15th rounds, Eddie, you lovingly looked into my eyes, put your hand on my shoulder and said, 'it's over, Son.' I*

was upset—I thought I could go on—but in your infinite wisdom
and compassion, you halted the fight. Now that I am older and
wiser, I see the wisdom of that decision. You loved me—and you
loved my family. You wanted me to live to enjoy my children and
grandchildren.

I wish I could be with you tonight, my friend. Even though
I am not there, know that the name Eddie Futch is forever
engraved in my heart.

Sincerely,

Smokin' Joe Frazier

Given the hostility that underscored their timeless fights, it
would have been cheering to think that Joe and Ali had come to
the same soft landing. But it was always hard to know. In their final
years, a reporter would come along, catch Joe in a sour mood, and
suddenly, the old grudge would be back in the headlines. People
who knew him for years told me they were sure that Frazier carried his animus for Ali to the grave, that he had been wounded so
deeply that he could never let it go. But others were certain that
he had come to peace with Ali, particularly those who knew him
from his boyhood days in South Carolina. Nearly all told me some
variation of "Billy never hated a soul in his life." No one can know
for certain, of course; any conclusion either way is perhaps clouded
by your own thoughts of how forgiving you would be in the same
circumstances. Probably, some of it also has to do with your own
belief in the power of reconciliation, how you define "unforgiv-
able," and the enchantment of happy endings. But then a story was
told to me by Darren Prince, CEO and president of the Prince
Marketing Group.

While it is fairly well known that Joe and Ali got together at
the NBA All-Star Game at the First Union Center in February
2002 and brokered a peace, only Prince and a handful of others

were aware of what had occurred the evening before. According to Prince, he received a call the day before the game from Ali representative Harlan Werner. Lonnie Ali had contacted Werner and asked to pass along an invitation to Joe, Marvis, and Prince for dinner at the Alis' hotel in Philadelphia. Remembering how Frazier had reacted to the offer to join Ali and his wife on the red carpet at the movie premiere, Prince was not optimistic. In fact, he was certain that Joe would say no when he told him of the invitation. But Frazier surprised him. "All right," he said. "I'll go see the Butterfly."

Joined by one of his associate agents, Nick Cordasco, Prince arrived at the hotel with Joe and Marvis and got on the elevator. As they were riding up, Cordasco looked down and gasped. He grabbed Prince by the arm and whispered, "Bro, Joe has a gun in his sock." When they got off the elevator, Prince pulled Marvis aside and said, "Marvis, what the fuck, Pop has a gun in his sock?" Marvis doubled over in laughter. He told Prince, "Pop got a license to carry that years ago from being harassed and having some issues. Nothin' to worry about." Prince remembered, "So that kind of loosened up the mood. And then we knocked on the door."

Lonnie opened it, gave Joe a hug, and led them inside the suite. Ali was sitting on a couch. Nearby were his longtime friend and photographer Howard Bingham and Bingham's son, Damon. With a big smile, Joe half trotted over to his old rival. "Having been around him, I could see it had been a rough day for Ali," said Prince. "Maybe it was his medication. He was bloated." But Prince would remember that Ali beamed when he saw Joe. He leaned forward and Joe helped him up. He grabbed Ali under his arm and Ali embraced him in the hug. "Like a baby, Muhammad stood there with his face on Joe's shoulder," said Prince. "Both of them had tears rolling down their faces." Prince remembered that Lonnie said, "Joe, thank you. Muhammad just found peace."

Plates of delicious food were ferried into the room and placed on the table by waiters. Ali sat at one end of the table, Joe at the other end. "There would be days when it was hard to watch Ali eat," said Prince. "He would drool and have to have a bib on. Lonnie would have to wipe his face." As Lonnie recommended to Joe a back specialist her husband had used—Joe had told her he had been experiencing some back pain—Ali looked at Frazier across the table, bit his bottom lip in a gesture of pretend ferocity and said:

"Joe Frazier.

"Joe Frazier.

"Joe Frazier.

"I want you. Hey, Gorilla."

Joe dropped his fork on his plate, looked him in the eye, and said, "Man, Butterfly. I kicked your ass for three goddamn fights. Are we going to have to go at it again?"

Prince remembered, "Muhammad was crying 'cause he was laughing so hard."

As the dinner came to a close, Marvis suggested they hold hands and pray. He stood between his father and Ali. As he began, his father stopped him and said, "Hey, son, I got this."

Joe then lowered his head and prayed: "Dear Lord. We have forgotten and we have forgiven. Please heal this man, because he has given so much to the world. And he has grandkids and babies, and it's time that he can really enjoy his life. So please do whatever you can to make this man right again."

At the all-star game the following day, the NBA seated them side by side at courtside. Photographers snapped their picture, which would appear in papers across the world. Celebrities who had been just children or not even born yet when the two of them were at the height of rivalry looked on in something close to reverence: Magic Johnson, Michael Jordan, Kobe Bryant, Jamie Foxx, Justin Timberlake, Britney Spears, Samuel L. Jackson, and on and

on. Alicia Keys sang "America the Beautiful," and as the crowd settled back into their seats at the end of the song, Ali and Frazier remained standing together. When the spotlight moved to them, the
crowd once again stood and began cheering. As the noise washed
over them, so loud and so long that it seemed to go on forever, Ali
turned to Joe and whispered in his ear.

"Hey, Champ."

"Yeah, Champ?"

"We're still two bad brothers, aren't we?"

"Yes, we are, man. Yes, we are."

Joe with three of his children (L to R): Jacqueline, Weatta, and Marvis, 1968.
Philadelphia Bulletin

ACKNOWLEDGMENTS AND SOURCES

By a coincidence that now somehow seems to me to have been oddly ordained, the first prizefight I ever saw in person was the rematch between Joe Frazier and Jerry Quarry on June 17, 1974, at Madison Square Garden. Then seventeen years old, I had come to New York as an end-of-the-school-year getaway to spend a week with my father, Mark Kram, who was then at the height of his career as a writer for *Sports Illustrated*. While he covered the brawl that unfolded from a seat at ringside, I sat thirty or so rows back and became swept up in the havoc Frazier wreaked upon poor Jerry, whose face let loose geysers of scarlet before the referee, Joe Louis, halted the fight. At the office with Dad the following day, I looked on from a chair behind a tower of books and old newspapers as he fed a piece of paper into his typewriter, lit the pipe he clenched between his teeth, and began writing. I remember thinking as the room became clouded with smoke that he attacked the keys with the same gusto with which Joe had imposed himself upon Quarry.

Dad was a formidable figure in the realm of sportswriting. At the Thrilla in Manila, in October 1975, he produced one of the finest event pieces in the annals of journalism. Near the end of his life, at age sixty-nine, he revisited the Ali-Frazier trilogy in his book, *Ghosts of Manila: The Fateful Blood Feud Between Muhammad Ali and Joe Frazier* (HarperCollins, 2001). The book was an essay-istic take on the world he knew so well, a deep blend of reporting and contrarian views that would elevate Frazier in the public eye. But *Ghosts of Manila* was not a biography, and a comprehensive one had still not been done when it occurred to me in 2016 to give it a whirl. By then, it had been just under five years since Joe had passed away and just under forty since he had faced Ali in Manila. Ten years from now or even five, it seemed to me that it would be impossible to do the book that I envisioned, one that coupled whatever remaining primary sources I could find with my own experiences with Joe during a twenty-six-year career as a sportswriter in Philadelphia.

This book would not have been possible without the generous help of Weatta Frazier-Collins, who has dedicated her life to keeping the spirit of her father alive through The Legacy Exists, a foundation that hands out scholarships to underprivileged children. I am thankful to Weatta and her husband, Gary, for their trust and support. I have interviewed members of the Frazier family including Martha (Mazie) Rhodan, Joe's last remaining sibling; Marvis Frazier; Joseph Jordan Frazier; Derek Dennis Frazier; Lisa Coakley; Dannette Frazier; Rodney Frazier; Mark Frazier; Miriam Frazier; Ollie Frazier; Annie Green; Frances Morrell; Vernell Williams; and Tom and Ginger Bolden. I am particularly grateful to Mazie, who sat down with me for a total of six hours of interviews and provided me with the written account that appears in chapter 1 about the day Joe was born.

Significant assistance also came from Denise Menz, who spoke candidly with me about her forty-odd-year relationship with Joe. From 1968 until Joe died in 2011, Denise was a behind-the-scenes fixture who had shied away from public attention. I would also like to thank Sharon Hatch and Sherri Gibson, both of whom also had long relationships with Joe and provided me with some valuable insights.

Over the two-year span during which I reported the book, I made two extended trips to Beaufort, South Carolina, including one for the 2017 Frazier family reunion. With the cheerful guidance of Dannette Frazier, I found my way to John Trask III, whose family had owned the commercial farm where Dolly Frazier worked and became an invaluable aid. Interviews in Beaufort also included John Trask Jr., Pastor Kenneth Doe, Isaac Mitchell, Matthew McAlhaney, and Lottie Antley. Historian Lawrence S. Rowland provided me with an overview of Beaufort County in both an extensive interview and in the book he coauthored with Stephen R. Wise, *Bridging the Sea Islands' Past and Present (1893–2006): The History of Beaufort County, South Carolina, Volume 3* (University of South Carolina Press, 2015). For a deeper understanding of the Gullah culture, I referred to the book by Roger Pinckney, *Blue Roots: African-American Folk Magic of the Gullah People* (Sandlapper Publishing Co., 2007). For background on day-to-day life in Beaufort in the 1940s and '50s, the visit by Joe Louis, and the Stinney and Feltwell cases, I appreciate the assistance I received at the Beaufort County Library, where I was also able to unearth old copies of the *Beaufort Gazette* on microfilm. The Beaufort County School System provided me with Joe's academic record.

Key interviews in Philadelphia and elsewhere included Kevin Dublin, who helped inform my portrayal of Hector Frazier; Philadelphia boxing promoter J Russell Peltz; the late boxing trainer George James, who shared with me his observations on Yank Durham, Sonny Liston, and Gypsy Joe Harris; the ex-PAL boxer Al Massey, who I interviewed at the Chester State Correctional Institution; Joe Hand, who was with Frazier during his Cloverlay years; Joe Hand Jr.; Frank Rizzo Jr.; Philadelphia disc jockey Jerry Blavat; Tad Dowd, who owned a piece of Oscar Bonavena; Jerry Ellis, whose brother Jimmy twice fought Frazier; the late Los Angeles matchmaker Don Chargin; promoter Don Elbaum; former Ali aide Gene Kilroy, who provided me with perspective on the feud between Ali and Frazier; Gloria Hochman, who shed light on the long relationship between Frazier and her late husband, Stan;

Gene Seymour, who analyzed the relationship between Chuck Stone, Frank Rizzo, and the City of Philadelphia; Les Pelemon, who helped Frazier launch his singing career and was with him in the days up to and including the Fight of the Century; ophthalmologist Dr. Myron Yanoff, who furnished me with a written account of his treatment of Frazier; publicist Bob Goodman, who shared his memories of the old Madison Square Garden and the row that occurred between Frazier and Ali on the Cosell set; Gordon Peterson, who worked publicity for promoter Don King and was attached to the Frazier camp in the weeks leading up to the Thrilla in Manila; Patti Dreifuss, who also worked publicity for that promotion; New Jersey boxing commissioner Larry Hazzard; heavyweight champion Larry Holmes; assistant trainer Val Colbert; the singer Michael Averona; Philadelphia boxing historian John DiSanto; former business manager Burt Watson; cardiologist Dr. Nicholas DePace; former promoter Joe Verne; photographer George Kalinsky; artist Richard Slone; Dr. John Kelly, the orthopedic surgeon who helped Frazier coordinate his cancer treatments; and Dr. Robert Cantu, clinical professor of neurology and cofounder of the Chronic Traumatic Encephalopathy (CTE) Center at the Boston University School of Medicine. Agent Darren Prince shared with me the tender encounter between Frazier and Ali at the 2002 NBA All-Star Game.

For my portrayal of Gypsy Joe Harris, I conducted an extensive interview with his brother, Anthony Molock, who authored the book *Gypsy Joe Harris: Son of Philadelphia* (AuthorHouse, 2006). For the argument that occurred between Frazier and Gypsy at the gym at the beginning of chapter 11 and other interactions between the two, I used the dialogue as Molock presented it. For the angry confrontation with Ali on his doorstep in Philadelphia, I used a trimmed-down version of what appeared in *Ghosts of Manila*, which was also confirmed and used by Molock in his book. Useful as well were the two profiles of Gypsy Joe that my father published in *Sports Illustrated*, in June 1967 and March 1969. For an overview of

Gypsy Joe in his later years, there was no better source than the profile written by Robert Seltzer for the *Philadelphia Inquirer* in February 1989.

For my portrayal of Eddie Futch, I augmented extensive interviews I had with him in Reno, Nevada, in 1986 and in the Poconos in 1993 with additional reporting from the journalist Sunni Khalid, who provided me with the transcripts of six hours of interviews he had conducted with Futch. I also spoke at length with Eva Futch, his widow, who gave me a copy of the congratulatory letter that Joe sent Eddie prior to being honored by the NAACP. Gary Smith also penned a probing profile of Futch in *Sports Illustrated*.

Some splendid writers chronicled the career of Joe Frazier. One of the delights of doing this book is that I became reacquainted with some of them on the page and in conversation. To a man, they were knowledgeable, irreverent, often hilarious, and there every day. I had the good fortune to work with one of them for years at the *Philadelphia Daily News*—Stan Hochman, who covered Joe from the infancy of his career with flair and scrupulous attention to the facts. Although Stan had passed away by the time I started this book, I conducted long interviews with some of his more distinguished colleagues on the beat, including Larry Merchant, who, as sports editor of the *Philadelphia Daily News*, reinvented the American sports pages into a place where journalism actually happened; the late Tom Cushman, a lovely writer who covered the Ali beat for the paper; the late *New York Times* columnist Dave Anderson, whose reporting skills were only surpassed by his generosity of spirit; *Newark Star-Ledger* columnist Jerry Izenberg, who knew Frazier and Ali as well as anyone; and Robert Lipsyte, the erudite former *New York Times* columnist who charmed me fifty years ago with the publication of his young adult novel *The Contender*.

Work that appeared in three of the big papers then in Philadelphia informed the historical context for the book. Along with Hochman and Cushman at the *Philadelphia Daily News*, I drew on the work of Will Bunch, Bill Conlin, Ed Conrad, Paul Domowitch, Ray Didinger, Bernard Fernandez, Thom Greer, Rich Hofmann, Jack McKinney, Larry McMullen, Bill Shefski, Elmer Smith, Leon Taylor, Gary Smith (later of *Sports Illustrated*), Chuck Stone, and Mark Whicker (whose columns appeared previously in the *Philadelphia Bulletin*). Coverage at the *Philadelphia Inquirer* included reporting by Gene Courtney, Frank Dolson, Lewis Freedman, Tom Fox, Hoag Levins, Jack Lloyd, Bill Lyon, Joe McGinniss, Skip Myslenski, Sandy Padwe, Samuel L. Singer, and Bill Thompson. Star columnist Sandy Grady served up the laughs over at the *Philadelphia Bulletin*, where he had a support staff that included Jim Barniak, Hugh Brown, Jack Fried, George Kiseda, Claude Lewis, Daniel J. McKenna, Lee Samuels, and Bob Wright. At *Philadelphia Magazine*, Maury Z. Levy chimed in with an illuminating profile of Ali.

From outside Philadelphia, an array of journalists weighed in on Joe and his opponents. Along with Dad at *Sports Illustrated:* Robert H. Boyle, Houston Horn, Martin Kane, Douglas S. Looney, Tex Maule, William Nack, Jack Olsen, Pat Putnam, Gilbert Rogin, Mort Sharnik, and Edwin Shrake. Along with Anderson and Lipsyte at the *New York Times:* Arthur Daley, Michael Katz, William C. Rhoden, Richard Sandomir, Red Smith, and Peter Wood. At the *Los Angeles Times:* Don Hafner and Jim Murray. At the Associated Press: Ed Schuyler Jr. At the *Washington Post:* Shirley Povich. At the *New York Post:* Milton Gross. At the New York *Daily News:* Phil Pepe, who also authored the early biography of Frazier *Come Out Smokin': Joe Frazier—The Champ Nobody Knew* (Coward, McCann & Geoghegan, 1972). At *Sport* magazine: Dick Schaap. At *Life* magazine: Thomas Thompson. At *Esquire*: Cal Fussman. *Time* magazine also provided a plethora of unbylined reporting.

Key articles consulted: Joe Flaherty profiled Sonny Liston for *Esquire* in March 1969. Flaherty also covered the Thrilla in Manila in a series for *The Village Voice*. At *The Saturday Evening Post*, Bruce Jay Freidman had a September 1967 profile of Joe that delved into his childhood in South Carolina. *Harper's Magazine* carried a piece by Perry Deane Young in February 1972 that was set at the Brewton Plantation and included a wealth of information on Dolly Frazier. *Playboy* magazine interviews with Cassius Clay, in October 1964; Joe, in March 1973; and George Foreman, in December 1995, provided insight into a wide range of topics; *Playboy* also ran a piece on Joe by Katherine Dunn in March 2012 that I found useful. Nik Cohn reported a compelling piece for *New York* magazine in October 1975 entitled "Ali, Racist," which I quoted from liberally in chapter 9.

Along with *Ghosts of Manila*, books that were helpful included *Smokin' Joe: The Autobiography*, by Joe Frazier with Phil Berger (Macmillan, 1996); *Only the Ring Was Square*, by Teddy Brenner as told to Barney Nagler (Prentice-Hall, 1981); *Once There Were Giants: The Golden Age of Heavyweight Boxing*, by Jerry Izenberg (Skyhorse Publishing, 2017); *Cornermen: Great Boxing Trainers*, by Ronald K. Fried (Four Walls, Eight Windows, 1991); *My View from the Corner: A Life in Boxing*, by Angelo Dundee with Bert Randolph Sugar (McGraw-Hill, 2008); *Inside the Ropes*, by Arthur Mercante with Phil Guarnieri (McBooks Press, 2006); *Muhammad Ali and the Greatest Heavyweight Generation*, by Tom Cushman (Southeast Missouri State University Press, 2009); *Facing Ali: 15 Fighters, 15 Stories*, by Stephen Brunt (Lyons Press, 2002); *Muhammad Ali: His Life and Times*, by Thomas Hauser (Simon & Schuster, 1991); *Champion: Joe Louis, Black Hero in White America*, by Chris Mead (Charles Scribner's Sons, 1985); and *Ali: A Life*, by Jonathan Eig (Houghton Mifflin Harcourt, 2017). For an overview of Frank Rizzo and his political climb, I used *The Cop Who Would Be King: The Honorable Frank Rizzo*, by Joseph R. Daughen and Peter Binzen (Little, Brown and Company, 1977). For an understanding of organized crime in Philadelphia, I drew on *Black Brothers, Inc.: The Violent Rise and Fall of Philadelphia's Black Mafia,* by Sean Patrick Griffin (Mile Books, 2007). I also sat down with Sean for a lengthy interview in Charleston, South Carolina.

For their advice and review of the manuscript, I owe a debt of gratitude to the expertise of Sunni Khalid, who also assisted me with his comprehensive analysis of the Ali and Foreman bouts; former *Daily News* colleagues Doug Darroch and Bernard Fernandez; David Borsvold; Dan Trigoboff; John DiSanto; and Tom Lachman, a talented former *Washington Post* editor who years ago gave me my first assignment on our college paper. Help along the way also came from George Bochetto, Dave Gambacorta, David Hanna, Gene Bonner, Ken Hissner, Glenn McCurdy, Trudy Menz, Michael Mercanti, Marc Steiner, Ashley Sylva, and Spencer Wertheimer. I am also appreciative of the staff at the Special Collections Research Center, Samuel L. Paley Library at Temple University and to Jackie Koenig for her assistance in the transcription of interviews.

For their encouragement through the years: John Schulian, who has not written a word that has not inspired me; Michael and Teresa Capuzzo; Stephen Carroll; Norman Chad; Joe and Dottie Distelheim; Mike Downey; Eliot Kaplan; Michael Leahy; Jeannie Leto; and Ron Rapoport. Chances are I would not have been in a position to do this book had I not found my way to Philadelphia in 1987. For that, I am thankful to former *Daily News* sports editors Mike Rathet, Brian Toolan, and Pat McLoone, along with two assignment editors who were enormously helpful, the late Jeff Samuels and Paul Vigna.

No one has been more supportive of me than Andrew Blauner, who is not just a talented literary agent but one of the true gentlemen of his profession. I could not have been more pleased when he placed the book at Ecco with Daniel Halpern, nor with the careful attention it has since received from editor Gabriella Doob.

To my family . . . thank you. My wife, Anne Johnson, has been as giving as any spouse could hope to expect and is as an astute an editor as any I have come across. No word goes out the door without it passing under her eyes. Our daughters, Cory and Olivia, have been an enduring source of pride for us both; this book is dedicated to them with the prayerful hope that they find their way in this troubled world. Love and peace to Greg and Tracey Franz and their children, Lily and Theo; Peter and Kerry Goldberg and their children, Emily, Ben, and Charlie; Rene; Alix and Matthew; Raymond and Hoi-Ling and their children, Dylan and Ava; and Robert and Heather.

A parting word: My mother, Joan, passed away during the writing of this book. Whenever I think of her—and it is often— I am reminded of something Joe once said: "When your mom dies, that's you." No one ever had a finer one.

INDEX